A PRACTICAL GUIDE TO
VALUE CLARIFICATION

DISCARD

A PRACTICAL GUIDE TO VALUE CLARIFICATION

by

MAURY SMITH
O.F.M., D. Min.

UNIVERSITY ASSOCIATES, INC.
7596 Eads Avenue
La Jolla, California 92037

I believe it will also be humbling for us if we admit freely that we are impelled to our interest in values not only by the intrinsic logic of science and philosophy but also by the current historical position of our culture, or rather of our whole species. Throughout history values have been discussed only when they became moot and questionable. Our situation is that the traditional value systems have all failed, at least for thoughtful people. Since it seems to be impossible for us to live without values to believe in and approve of, we are now in the process of casting about in a new direction, namely, the scientific one. We are trying the new experiment of differentiating value-as-fact from value-as-wish, hoping thereby to discover values that we can believe in because they are true rather than because they are gratifying illusions.

Maslow, *The Farther Reaches of Human Nature*

From Abraham Maslow, *The Farther Reaches of Human Nature*, p. 151, © 1971 by The Viking Press. Used with permission.

PREFACE

Getting involved in value clarification has been an exciting and growthful experience for me.

My approach to value clarification has been influenced not only by the pioneers in the field—Raths, Simon, and Kirschenbaum—but also by my experience with psychosynthesis and its emphasis on the will and transpersonal psychology.

There is a great deal of similarity between value clarification's "choosing" and Assagioli's "act of will." About the act of will, Assagioli (1965) says,

> Since the outcome of successful willing is the satisfaction of one's needs, we can see that the act of will is essentially joyous. And the realization of the self, or more exactly of being a self (whose most intrinsic function, as we have seen, is that of willing), gives a sense of freedom, of power, of mastery which is profoundly joyous. (p. 201)[1]

This sense of freedom, power, and mastery is also the objective of value clarification.

The transpersonal or spiritual aspect of man usually involves theology, morals, faith, etc., which emphasize a value stance and thus are not listed as criteria in the process of valuing. I conceptualize man as a total, unified person; I include the spiritual component because I believe firmly that man's human and spiritual natures are united. One does not have to accept this definition in order to use or accept the theory of value clarification, but it is important for me to state my own values and biases to the reader.

My thanks and appreciation are extended to Brian Hall for introducing me to value clarification; to J. William Pfeiffer, president of University Associates, for his invitation to write this book, for his consultation during the writing process, and most of all for the friendship we have enjoyed for the last several years; and to Arlette Ballew,

---

[1]Roberto Assagioli, *Psychosynthesis,* © 1965 by the Viking Press. Used with permission.

Anthony Banet, and Barbara Rose at University Associates for polishing and refining the text. I would like to express an especially deep gratitude to my Franciscan Community (Anton Braun, Joseph Jansen, Elric Sampson, Joseph Sandlin, Paul Schullian, and Martin Wolter) who have at times allowed me to experiment on them with value-clarification activities but who, most of all, have supported and encouraged me in many ways. Jean Wawrzyniak, our secretary, did a magnificent job of protecting me from the telephone and of helping out in emergencies so that I would have the time to write.

Maury Smith
Indianapolis, Ind.
September, 1976

TABLE OF CONTENTS

INTRODUCTION

The goal of this book is to provide a facilitator with sufficient knowledge to be able to create his own value-clarification designs. There are six major segments to the book:

Chapter 1 presents an explanation of the basic theory of value clarification; Chapter 2 comprises twenty-nine activities with variations, which may be used by the facilitator in adapting and designing new approaches; Chapter 3 discusses design considerations relevant to value-clarification programs and gives a check list for evaluating new activities and designs; at the end of Chapter 3, sample designs are provided for an introductory opening session, one-day and three-day sessions, and a series of eight sessions, to help a new facilitator get started; and Chapter 4 includes background readings in value clarification. The concluding section is an annotated bibliography, which is provided as a resource for those who may want to do additional reading and studying in the field.

The process suggested to a new value-clarification facilitator is the following: 1) do the activity yourself and take process notes on your reactions to the experience; 2) try the activity first as an experiment with a small group, then with a larger group; 3) adapt the technique to the group you will be working with—create your own activity; and 4) keep process notes on the reactions of the group.

The book is intended for educators, counselors, group facilitators, organization development consultants, and those involved in various forms of human relations training, experiential learning, and laboratory education. Every attempt has been made to present the information simply and clearly so that those who are beginning as facilitators will understand the process. Those who are experienced in group work will be able to move more rapidly in creating their own designs. In general, both the beginning and the experienced facilitator will begin with the educational level of intervention, move to the engaging level of intervention, and then, when comfortable in so doing, use the confronting level of intervention. (This process is fully explained in Chapter 3.)

1

Value clarification is a simple but powerful process that can help increase their awareness of a vital area of life. And to that end, this book is offered as a guide to those who value helping others to become all they can be.

CHAPTER 1

THE NEED FOR EXPLORING VALUES

"What is life all about?" "Where do I fit in?" "How can I find meaning?" These and similar questions, which seek answers about values, are being heard with increasing frequency in contemporary society. Today, people have greater mobility, more variety, a wider range of opportunities, and more potential for change than ever before in human history. It is no longer uncommon for people to change careers (or even spouses) three or four times in a lifetime; our choice of leisure activities is limited only by a lack of financial resources or a lack of imagination; and close relationships can be maintained all over the globe—with the assistance of instant telecommunications and jet travel. Yet, we pay a price for these luxuries that now seem so commonplace. That price is paid in the form of having to choose who and what we will be. We must make decisions about issues that were never before even called into question. Only a generation or two ago, children grew up knowing, to a far greater extent than we, what roles they would play in adult life, where they would live, and what they would believe in. These issues were decided before they were born. Today, though, we must not only make those decisions for ourselves, we must often re-evaluate our choices every couple of years. And it is for this reason that people are asking more value questions. In order to begin to find answers to these questions, we need first of all to be able to identify what values are. To compound the difficulty of answering value questions, we must contend with rapidly changing cultural norms that deeply affect the value systems of individuals and societies.

CHANGING VALUES

We are not born with values, but we are born into cultures and societies that promote, teach, and impart their values to us. The process of acquiring values begins at birth. But it is not a static process. Our

3

values change continually throughout our lives. For example, as children, our highest value might have been play; as adolescents, perhaps it was peer relationships; as young adults, it may have been relationships with the opposite sex; and as adults, our highest value may be the work we do. For many older people, service to others is the highest value.

Because values seem to change so rapidly, people are often confused about the reality or validity of the values they hold at any one time. Individuals need to discover why and how they have developed the values they have.

VALUES DEVELOP FROM LIFE EXPERIENCES

Young children tend to adopt the values of those who most influence them: parents, guardians, teachers, clergymen, and so on. As the child becomes an adolescent and young adult, he begins to explore and question the values he took from others and starts to make choices about which values he wants to assimilate and affirm and which he finds intrusive. Relationships that foster dependency promote situations in which values are introjected, or imposed, rather than consciously and freely chosen. Thus, when an individual is in a dependent relationship, he may adopt a particular set of values in order to maintain that relationship. Here, the exploration and questioning process is hindered by a desire to please and be pleasing, and the young adult forestalls the opportunity to exercise his own judgment about the values he holds.

Books, movies, television, travel, neighborhoods, schools, friends —all influence the formation of our values. We are formed largely by the experiences we have, and our values form, grow, and change accordingly.

VALUE CLARIFICATION:
A PRACTICAL APPROACH

The pluralism, transiency, and mobility of contemporary life styles affect people emotionally; the necessity of quickly making many decisions creates tension and stress. This theme has been ably and thoroughly explored by two prominent theorists in the field of psychotherapy: Rollo May in *Love and Will* and Roberto Assagioli in *The Act of Will*. These writers point out the necessity of learning how to make decisions, set goals, plan one's life, and develop ideals that give meaning to life. Gordon Allport, Carl Rogers, and Abraham Maslow are other psychologists who have made important contributions in this area. More recently, Louis Raths, Merrill Harmin, and Sidney

Simon have developed an approach to education that utilizes value questions—an approach that has come to be called *value clarification.*

The Method

In the past, philosophy and theology sought to understand and define values: objective, ontological, metaphysical, and moral values. Most of us still feel the effects of the Puritan and Victorian eras, when values were defined primarily in terms of moralistic "shoulds" and "should nots." Value clarification as a methodology considers this moralistic stance to be an imposition upon the individual of predetermined values, and it seeks instead a method whereby individuals can discover their own values. Thus, value clarification does not tell a person what his values should be or what values he should live by; it simply provides the means for him to discover what values he does live by.

The Focus

To most people the word "value" carries many vague and diffuse meanings, and it is used almost indiscriminately in the language. Even in the academic worlds of philosophy, psychology, and theology, there are about as many definitions of values as there are people who think about them. Most of these academics are concerned with an intellectual definition and think in terms of absolute values, but this approach is not suitable for the purposes of this book. Our aim here is simply to define value clarification and, later, to provide techniques whereby individuals can discover what values they hold. Raths, Harmin, and Simon (1966) describe the value-clarification approach as follows:

> We shall be less concerned with the particular value outcomes of any one person's experiences than we will with the process that he uses to obtain his values. ... We do, however, have some ideas about what *processes* might be most effective for obtaining values. These ideas grow from the assumption that whatever values one obtains should work as effectively as possible to relate one to his world in a satisfying and intelligent way. (p. 28)

To avoid an imposition of preconceived values, value clarification focuses exclusively on the *process* of valuing (how people come to the values they have). Kirschenbaum (1976) defines the valuing process as "a process by which we increase the likelihood that our living in general or a decision in particular will, first, have positive value for us, and second, be constructive in the social context" (p. 102). (Because "process" has become a commonly used term in the behavioral sciences—especially in social psychology, organization development, and conjoint and group therapy—it is useful to note here that, in the context of this volume, "process" means *how a system is doing what it is doing*.)

In the process of clarifying one's values, a discrepancy often arises between what a person *says* his values are and how he is actually *behaving*. The value-clarification process demands that a person relate to himself, to others, to society, and to his environment, but whether and how the person unifies his thoughts, feelings, and actions is his own decision. Through the use of value-clarification activities, individuals are invited to initiate an exploration of their own valuing processes, and within the context of a group situation, they are encouraged to use one another as resources. It is hoped that this process will facilitate a congruence between one's thoughts and one's actions.

The Applications

Value clarification is useful in:

- personal growth and development of identity;
- staff development and personal goal setting within an organizational context;
- psychotherapy, guidance, and counseling, to integrate thoughts, feelings, and actions;
- education, to stimulate an active search for relevance;
- organizations, for team building and conflict resolution; and
- religious organizations, to discover spiritual guidelines.

To explore one's values is to become more thoroughly aware of oneself and, consequently, to live in closer harmony with one's beliefs. Value clarification helps develop a person's (or group's) ability to make decisions about the direction of his life (or organization). Personal growth and a deeper commitment to one's life goals is achieved through an increased understanding of how to make choices among the myriad of options available. But value clarification is *not* a panacea. It is one of many approaches to growth and development. It is not meant to replace psychotherapy, organization development, and other approaches; however, it may be viewed as a powerful adjunct to these human relations disciplines. Through an awareness of the process of valuing, people can promote the change they want, and through the discovery of values, they can learn what it is they want.

PRIMARY VALUES

It is my view that primary values are so essential to the growth of man that they merit specific explanation in themselves. By primary values are understood values that are chosen, acted upon, that a person is happy with, that are necessary for the authentic development of man. (Hall, 1973b, p. 106)

Primary values, then, are basic; they are prerequisite to natural human development. Hall (1973) has postulated two primary values: *self-value*, which is "the ability to accept that 'I am of total worth to significant others,'" and *value of others*, which is defined as the knowledge that "others are of total worth, as I am." In other words, I value myself and I value others.

These primary values cannot exist except in relationship to each other. I cannot affirm others unless I affirm myself, and unless I affirm myself, I will not be able to affirm others. The implication of this aphorism is that value clarification promotes respect, understanding, and concern for self and others.

FULL VALUES

From their study of social and educational psychology, and as a result of their observation of the process of forming values, Raths, Harmin, and Simon (1966) identified seven criteria which, together, constitute an operational definition of a full value. This definition is used in value clarification as the yardstick for determining whether a person is actually holding a value or not:

CHOOSING: 1) freely
             2) from alternatives
             3) after thoughtful consideration of the consequences of each alternative
PRIZING: 4) cherishing, being happy with the choice
             5) willing to affirm the choice publicly
ACTING: 6) doing something with the choice
             7) repeatedly, in some pattern of life. (p. 30)

These seven criteria should be considered in relation to an individual's strongest value and tested against his own definition of a value. Any criterion may be rejected or any other may be added. An eighth criterion—enhancing—is suggested and explained later in this chapter. Values are not formed by impulse, thoughtless action, following the crowd, or blind acceptance of others' values. Full values are formed by a process that involves one's feelings, thoughts, desires, actions, and spiritual needs. It is a dynamic formulation, not a static one.

The first three criteria, those of choosing, rely on the individual's cognitive abilities; the second two criteria, prizing, emphasize the emotional, or feeling, level; and the last two criteria, acting, are concerned with external behavior. It can be seen that values are formed by a combining of all the features that constitute our humanness: intellect, will, emotions, and spiritual needs. And although it may seem, at times, that we can distinguish between these parts of ourselves, in reality they cannot be separated. Who we are and what we do derives from the dynamic and constant interplay of these forces.

First Criterion: A Value Must Be Chosen Freely

A full value is a guide, a norm, a principle by which a person lives. The values that a person chooses freely are the ones that he will internalize, cherish, and allow to guide his life. Free choice excludes overt coercion by tyrants and subtle, or gentle, coercion by loved ones. Introjected values from childhood are not full values. Physical or environmental circumstances and societal laws may impose a value on a person that he does not espouse—to that degree it is not his value.

We have all experienced some form of hypocrisy in forming our own and attempting to form others' values—pretending to believe in something in order to make a good impression on someone else or preaching a value to others that we do not hold for ourselves. At home, at work, and in our social relations, all of us have undoubtedly experienced such instances; often, our responsibilities as parents, children, teachers, students, employers, employees, etc., require us to act or speak differently from the way we believe. It is well to remember, both for ourselves and about others, that full values are only those that are freely chosen.

Second Criterion: A Value Must Be Chosen From Alternatives

That a value must be chosen from alternatives follows from the first criterion that a value must be freely chosen. If there are no alternatives, there is no freedom of choice. In many situations, it may initially seem that there are alternatives when, in fact, subsequent evaluation shows that there are no alternatives. A thief demanding your money or your life is not really extending alternatives; nor is a parent who demands that a child eat his spinach or go straight to bed.

When persons, institutions, limitations of self or of the physical world impose the alternatives, they almost always remove the possibility of the person freely choosing from among the alternatives. For example, a person cannot value breathing, of itself, because there is no choice involved: a person must breathe to live. One can, however, value mountain air, or sea air, or a special breathing technique, such as Yoga.

At work, a person may be offered a choice between a number of undesirable tasks (according to that person's appraisal) and choose one of them only because he values supporting his family, not because he values that task above the others. On more important levels: a man may stay married, not because he values his wife, but because he values the institution of marriage; a woman may choose a particular career, not because she values that kind of work, but because she values the status and financial rewards that go along with it.

People in positions of authority often believe they are offering genuine alternatives or options when in reality they are not. For example, a teacher may give her students the choice of studying the Civil War either by writing a book report, attending a lecture, or participating in a discussion; an employer or manager may pride herself on always giving her subordinates an opportunity to choose among certain tasks. But if the teacher and the manager do not allow their students and subordinates to help in establishing the alternatives, they are imposing values and precluding an opportunity for free choice. In other words, there may have been alternatives that were not presented that the students/employees would have preferred.

Third Criterion: A Value Must Be Chosen After Considering the Consequences

More precisely, a value must be freely chosen after careful study of the consequences of each alternative. That is, the consequences must be known. If a person does not realize the consequences of a particular alternative, he does not know what is going to happen; he has therefore not freely chosen that consequence. For example, a person may choose one career over another because he thinks it will give him more freedom to exercise his value of creativity; he may not fully realize, however, that the greater responsibility that goes with his chosen career will diminish his creativity through tension and worry. It often happens that when choosing one value a person fails to realize the consequences to another value, which may be even more important to him. For example, a man may value, first, his family life and, second, having interesting work. Without careful consideration of the consequences to his values, he may take a job that requires a great deal of travel. Although traveling may make the job interesting, it also takes him away from his family. In this situation, the person inadvertently reversed his values so that an interesting job ranked higher than his family. Had the man been aware of this consequence, he probably would have reconsidered his choice.

Only after the foreseeable alternatives or options open to him are fully and clearly understood is a person able to make a free and intelligent choice. Many times, of course, the consequences of one's choices cannot be known in advance. This fact does not necessarily mean that a free choice has not been made; it does mean, however, that once the consequences are understood the person must re-evaluate his choice in light of the new information.

Fourth Criterion: A Value Must Be Performed

A value is acted upon, performed, carried out: it influences a person's behavior in some way. Thus, what a person does reflects his values. We tend to read literature that supports our values; we join clubs or informal groups whose members share our values and whose goals correspond to our values; and we even use money according to our values— for instance, if we value thrift, we save money; if we value travel, fine food, or private education, we spend a good deal of it.

The importance of a particular value may be judged in terms of how much time we spend on it. If a person professes a high value for political awareness, for example, but never reads a newspaper or magazine, and never votes or attends political meetings, it might be said that political awareness is not really his value.

A person will also expend more energy on his values. A woman who values cooking but who has a full-time job may stay up late at night or get up early in the morning to bake bread or to prepare a fine meal for her family or friends.

Because people act according to their values, their values give direction to their lives. If a person does not act on his proclaimed values, then what he is talking about may be a desire, a feeling, or an idea; it is not a value. Many of the activities and experiences that are part of the value-clarification process are designed to help people discover what things they act on, as opposed to what they simply desire, feel, or think. It is common for persons who are well educated to believe that they have many values that in reality they do not.

There is a difference between "thinking about a value" and "reacting to a value." Thinking about a value may very well be an indication that the particular value is being formed and being readied for action, especially if a substantial amount of time is spent in planning. But a person who thinks of the particular value only when it is brought up by another person or mentioned in a book or on television, etc., is simply reacting to a value. In this case, the person merely agrees with something that has already been said or written. Reaction agreement without action may be a first small step toward the beginning of the formation of a value, but certainly it is one of the weakest value indicators possible.

Fifth Criterion: A Value Becomes a Pattern of Life

Values are acted on repeatedly and become life patterns. And the stronger the value, the more it influences one's life. For example, a person who strongly values service to others may choose to become a social worker. He may spend a great deal of time, energy, and money earning the necessary credentials, and he may even move his home to a remote area of the world where his service is most needed. His whole life becomes ordered around service to others. Although this example may be somewhat extreme, it is useful for understanding the extent to which a strongly held value can influence one's life. It also helps explain the motivation behind certain extreme actions. In the context of this criterion for a full value, it is worthwhile to look at the value of work in contemporary American society. The motivation for this value lies in the Protestant work ethic, which maintains that work brings salvation—union with God. People who hold this value get up early to go to work; because they abhor wasting time, relaxation breaks are short and meals are strictly functional (as opposed to other societies in which meals are long and provide an opportunity to relate to others). Work is valued over family life; many people bring work home and continue working during what is family time in other societies. This "homework" may be actual work or it may be time spent worrying about work and about whether one is advancing quickly enough. Work is valued over recreation, so additional work is done during what could be a time of leisure; often it is rationalized that this is a "different kind" or work, so it is "really" recreation. Many people in our society hold two jobs, and in many marriages, both spouses work. Quite a few people might be surprised to discover that they actually work eighty to ninety hours a week, if actual job time, homework, recuperation time, and worrying-about-work time are considered.

A value that becomes a pattern of life manifests itself in all aspects of one's existence: in dress, in friends selected, in the place one lives, in recreation time, in what one reads, in one's career, in the selection of a spouse, and in the way one relates to one's relatives.

Because the nature of the valuing process is dynamic, frequently a person will think that he still has a value that, in reality, he no longer holds. He assumes that because he once held a particular value he continues to hold it, and it may come as a great surprise when he finds it is no longer a value. For instance, a person may have once valued reading and spent several hours a day engrossed in books and magazines. If he is asked when was the last time he read a book and his response is "Three years ago," reading is obviously no longer a value in this person's life. In brief, a value tends to permeate and influence all aspects of one's life.

Sixth Criterion: A Value Is Cherished

A value is something a person feels positive about; he prizes it, cherishes it, respects it, rejoices in it, and celebrates it. As the individual grows toward full development of his values, he derives increasingly greater contentment, satisfaction, fulfillment, and joy from the act of choosing his own destiny.

Once a choice has been experienced as worthwhile and happy, it may be an indication of a value. If a person is not happy with the consequences of his choice, or if in the experience of the choice the person discovers that this is the wrong choice for him, it is not a full value. For example, a woman may hold a particular job, do well at it, and spend a great deal of time and energy on her work. But if the only reason she maintains the job is because it is the best she can do to support her family, and if she is not really happy with that particular work, then the job does not represent a value for her. The value is supporting her family, not the work she is doing to accomplish the value.

Another illustration comes from people who do volunteer work for community organizations. A woman may be asked to teach an evening class to members of the community. She may do it out of a sense of duty to her neighbors, but if she does not truly enjoy teaching, then the teaching is not a value for her. The value in this instance may be civic responsibility, service to others, or even the subject matter she is teaching, but it is not the act of teaching.

People cherish whatever helps them lead satisfying and productive lives. A person may perform certain activities for a while because of some other value, but if the activity itself is not cherished, it is not a value, and eventually the person may seek to remove that activity from his life—usually because it interferes with one of his values, such as the value of not having to perform unpleasant activities.

Seventh Criterion: A Value Is Publicly Affirmed

This criterion is directly related to the preceding criterion—that a value is cherished. When we have good news, we like to share it. When we discover a value that is freely chosen, the consequences of which we know, and that makes us happy, we want to tell others about it. In fact, if the value is a full value, we may even crusade for it. If we value a particular political ideology, we may campaign for the politician who holds the same value. A teacher who values a certain approach to education may promote seminars and encourage other teachers to adopt the same approach. An executive who comes to value a certain way of relating to subordinates will encourage his managers to adopt the same approach. In fact, the danger is that the person may be so enthusiastic about his value that he imposes it on other persons— causing them to "adopt" his value only as a matter of expediency. At the other extreme is the unfortunate person who is so apathetic about his life that he has few or even no values that he is willing to publicly affirm.

A possible exception to this criterion, however, might be a situation in which one's value conflicts with the norms of the society in which one lives. In this case, to publicly affirm his value, the person might be putting himself in jeopardy. A good example would be a Zionist who is living in Russia, or a Communist who is living in a small, politically conservative community in the United States.

Publicly affirming one's value has direct bearing on an eighth criterion, enhancement of the person's total growth. Generally, values help us to grow toward becoming what we consider a "good" person. If something moves us toward what we consider "bad," or unhealthy, we tend to hide it. If we are not growing by what we are doing, and therefore unwilling to make our position or activity known, then what we are doing is not really a value for us.

Eighth Criterion: A Value Enhances the Person's Total Growth

This final criterion is not so much a separate measurement of a full value as it is a natural outgrowth of the preceding criteria. In other words, if a value has been affirmed as a full value by having met the seven preceding criteria, it follows as a matter of course that that value will contribute to and enhance the person's total growth toward the goals and ideals that he has chosen for himself.

The prizing criteria (cherishing, affirming, enhancing) emphasize the emotional component of the process of valuing. We may choose and act, but if we do not also prize something, then it is not a full value. We are more apt to continue a full value as a pattern of life and act on it repeatedly when we prize it and find that doing it helps us to grow as a total person. People who dedicate themselves to their values experience joy regardless of the problems, difficulties, or sacrifices they may have to make. People who lack the quality of prizing in their lives lack vitality, tend to be apathetic, find life meaningless, and often suffer chronic depression.

Values are, by their nature, goal directed. As mentioned earlier, we tend to choose activities that relate to the values that we have selected as goals or even as ideals. In turn, our activities support or, in some cases, undermine our values. Value clarification does not attempt to advocate a particular direction for growth, it merely is a process by which a person can discover meaning and direction for his life through recognition of his values. Thus, if a value does not enhance the person, it is lacking one of the above criteria and is not a full value.

Summary: Criteria for the Process of Valuing

A full value is something freely chosen from alternatives after thoughtful consideration of the consequences of each alternative; it is acted upon repeatedly so as to become a pattern of life; it is found to give direction and meaning to life in such a way that it enhances the growth of the total person; and it is cherished and publicly affirmed to others. The emphasis in this definition is on the total person and the process of valuing. The definition comprises eight criteria, which can be used to assess whether a belief, attitude, or the like is truly a value.

The major areas of the above definition (choosing, acting, and prizing) correlate to the thinking, willing, and feeling aspects of the total person. The eight criteria point out the extent to which a value has been formed and developed; they do not themselves attempt to form or develop values.

KEY TO TERMINOLOGY

For increased clarity, the following chart attempts to equate terminology (as nearly as possible) across four areas:

COMMON USAGE	VALUE CLARIFICATION	PSYCHOLOGY	THEOLOGY/ PHILOSOPHY
thinking	choosing	cognitive	intellect
feeling	prizing	affective	emotions
behaving	acting	behavioral	volition
growing	enhancing	symbolic	spirit

FULL VALUES VERSUS PARTIAL VALUES

Any idea or activity that fulfills all eight criteria is, according to the value-clarification approach, a full value. If some, but not all eight, criteria are fulfilled, the idea or activity is a partial value. A partial value may be developing toward a full value or it may be diminishing. The mathematical possibilities of the combination of these criteria are endless. However, one should be warned against computerizing these criteria as if they were things or as if they could be measured in quantity. In value clarification, there is no effort to measure values. In dealing with quality and with meaning, one must simply accept the inevitable disparity between the two. Most of the time, when we refer to a value, we are talking about a partial value. For example, a person may claim that he values equal rights. He has freely chosen that value, has considered the consequences of the alternatives, cherishes and publicly affirms that value, and believes that it enhances his life. But he does not campaign for election of equal-rights political candidates, does not join activist groups or organizations, and does not spend much of his time engaged in related activities. In this case, it may be said that for this person, equal rights is a belief or aspiration—a partial value.

VALUE INDICATORS

A value indicator is an expression that points toward a value but does not necessarily fulfill all eight criteria of the process of valuing (Raths, Harmin, & Simon, 1966). Our full values grow out of and develop from these value indicators. But it is important to realize that not everything we do has to be, or should become, a full value. Humans are not perfect beings—we have complicated needs and desires which often conflict with one another, and as a consequence, only some of our values can become full values: we need only a few in order to grow and develop our whole potential.

Some of the more important value indicators are goals, purposes, aspirations, attitudes, interests, feelings, beliefs, convictions, activities, worries, problems, daydreams, use of time, use of money, use of energy, etc. (Rokeach, 1973, pp. 17-24 and 95-121). One may well substitute or add other value indicators, according to his or her particular biases.

The relevance of these value indicators is that they help us to discover the values we are developing. For most people, value clarification happens within the areas of their value indicators. That is, most of the value-clarification techniques help individuals to focus on their value indicators. Many of the techniques call attention to the value indicators so that the person can estimate whether something is truly a value for him.

Because most people confuse a value with one of these value indicators, it should be pointed out that a goal, aspiration, belief, activity, etc., is not a full value in and of itself. For example, a person may have decided on a goal but not have acted on it. A person may have aspirations, but he has not considered, and would not choose, the consequences. A person may have a certain belief and discover he really has not freely chosen it and does not really cherish it. It can be seen that these are partial values, but certainly not full values, according to the eight criteria.

VALUES AND CHANGE: NEW DIRECTIONS

The value-clarification process is exciting for the group facilitator because he can directly witness the process of human growth and development. The facilitator's skills and awareness contribute to, and often create, an environment in which the person is free to develop and explore the values he wants to live by and let his life be guided by.

It has been found that simple exposure to the value-clarification process causes change in people (similar to the observer-effect phenomenon noticed by anthropologists studying cultures in the field). Rokeach (1973) contends that change takes place when the person becomes disturbed emotionally as a result of realizing the existence of certain contradictions in his value system. In order to eliminate the inevitable self-dissatisfaction that follows such a realization, the person will realign his values to coincide with his new self-conception.

> ... the basic mechanism that initiates a process of change is an affective state of self-dissatisfaction, which is induced when a person becomes aware of certain contradictions in his total belief system. The more such contradictions implicate self-conceptions, the more likely that they will induce self-dissatisfaction and the more likely that the ensuing changes will endure. Self-dissatisfaction will not arise if such contradictions do not exist or do not become apparent or, should they become apparent, they are denied or repressed. But if a person perceives such contradictions within himself as credible, his perception should generate self-dissatisfaction. To reduce or eliminate such self-dissatisfaction, a person will often find it necessary to realign values with self-conceptions. Value change should in turn lead to a cognitive reorganization of the remaining values in the system and changes in functionally related attitudes, and it should culminate in behavioral change. (from *The Nature of Human Values,* 1973, p. 286)[1]

Rokeach believes that his approach, in which change centers around the creation of values, is more effective than conventional approaches, in which change centers around strictly cognitive or behavioral learnings. He emphasizes that self-confrontation in regard to value creation can lead to lasting changes in personality traits, thus provoking a change in one's total personality structure. However, a good deal more research is needed before Rokeach's theory can be verified.

---

[1]Reproduced from M. Rokeach, *The Nature of Human Values.* New York: Macmillan, 1973. Used with permission.

Traditionally, those involved in the field of value clarification have been careful to avoid instances in which an imposition of values is a possibility. They have felt that any attempt to impose definitions, judgments, or directions for change is highly inappropriate. Rather, they have focused exclusively on the process of valuing, on the awareness of one's own values and value indicators, and on the self-evaluation and change that are outgrowths of that process.[2] Rokeach, now, takes this process a step further, suggesting not only that the valuing process itself creates change, but that there are differences among particular values. Rokeach has established a list of eighteen instrumental and eighteen terminal values, which can be defined and ranked according to statistical measurements. Another approach to value clarification is presented by Howard Kirschenbaum, who defines a value strictly in terms of "process." Kirschenbaum dislikes the notion of "criteria" and feels that Raths' seven criteria should not be considered absolutes. He thinks, rather, that the evaluation of a value should be limited to personal processes. Kirschenbaum therefore expands Raths' three major areas of valuing (choosing, acting, prizing) into five major processes: feeling, thinking, communicating, choosing, acting. The following outline represents Kirschenbaum's approach to value clarification (Simon & Kirschenbaum, 1973).

The Valuing Process
I. Feeling
   1. Being open to one's inner experience.
      a. awareness of one's inner experience
      b. acceptance of one's inner experience
II. Thinking
   1. Thinking on all seven levels.
      a. memory
      b. translation
      c. application
      d. interpretation
      e. analysis
      f. synthesis
      g. evaluation
   2. Critical thinking.
      a. distinguishing fact from opinion
      b. distinguishing supported from unsupported arguments
      c. analyzing propaganda, stereotypes, etc.
   3. Logical thinking (logic).
   4. Creative thinking.

---

[2]Value clarification is not value free, as Kirschenbaum (1976) points out. Creativity, autonomy, equality, and justice are some of the values embedded in the process.

     5. Fundamental cognitive skills.
       a. language use
       b. mathematical skills
       c. research skills

III. Communicating—Verbally and Nonverbally
    1. Sending clear messages.
    2. Empathic listening.
    3. Drawing out.
    4. Asking clarifying questions.
    5. Giving and receiving feedback.
    6. Conflict resolution.

IV. Choosing
    1. Generating and considering alternatives.
    2. Thoughtfully considering consequences, pros and cons.
    3. Choosing strategically.
       a. goal setting
       b. data gathering
       c. problem solving
       d. planning
    4. Choosing freely.

V. Acting
    1. Acting with repetition.
    2. Acting with a pattern and consistency.
    3. Acting skillfully, competently. (pp. 105-106)

Kirschenbaum's description of the process of valuing and his emphasis on respect for the individual's discovery and control of his own processes is quite detailed; however, his schema may unnecessarily fragment the valuing process. The term "process" refers to the total dynamic procedure whereby values are developed and brought to fruition. "Criteria," on the other hand, are the standards by which it can be determined whether a perceived value is a true value. The term "criteria," therefore, implies that a particular aspect of the valuing process has been assimilated and actualized.

The functional definition of a value must be integrated into an understanding of the individual as a total living being. Thus, the emphasis in this book is on the unity between choosing and thinking, acting and willing, prizing and feeling, and enhancing and growing. In a unified version of the individual, this conceptualization also considers the human tendency toward self-actualization as manifested in behavioral and spiritual growth.

CHAPTER 2

VALUE-CLARIFICATION
TECHNIQUES

The "technology" of value clarification consists of a variety of structured experiences: fantasies, role plays, check lists, drawings, small-group tasks, and problem-solving activities. Some of these structured experiences originated in classrooms, clinics, or workshops; some were designed by the pioneers in humanistic education; some were adapted from classic textbooks; and still others are modifications of standard activities in human relations training. Most teaching and training techniques can be incorporated into a value-clarification program.

The twenty-nine structured experiences presented here represent the most popular and widely used value-clarification strategies. For the convenience of the group facilitator, they are arranged in the format designed by J. William Pfeiffer and John E. Jones for their *Handbook of Structured Experiences for Human Relations Training*. Where special materials are required, they are so noted in the format. All activities presuppose a physical environment conducive to large- and small-group experiences (i.e., a large room with usable wall space and movable chairs).

Many of these activities appear in some of the major publications in the field, including *Values and Teaching* by Raths, Harmin, and Simon; *Values Clarification: A Handbook of Practical Strategies for Teachers and Students* by Simon, Howe, and Kirschenbaum; and *Meeting Yourself Halfway* by Sidney Simon. Credit references are given for each structured experience; in each instance, an effort has been made to determine the originator of the activity or its prior publication in another form.

1. VALUE LOVE LIST

Goals

I. To explore how value indicators may be used to discern values.

II. To discuss the functional definition of a value.

III. To introduce a group to the value-clarification process.

Group Size

Any size group.

Time Required

One-half hour for completion of the listing, with additional time as needed for sharing in smaller groups.

Materials

I. Paper and pencils for all participants.

II. Newsprint and felt-tipped pens.

III. Worksheet I.

Physical Setting

A large room with movable chairs.

Process

I. The facilitator explains the goals of the activity. He instructs the participants to make a list of the ten activities they love to do most and explains that these ten items may include anything the participants can think of. (Formulating this list may take from four to eight minutes, depending on the size and speed of the group.)

II. Participants are told to rank, on the left side of the list, their items from one to ten, according to how much they value each item. "1" indicates the most valued, "2" the next most valued, etc.

III. Using newsprint, the facilitator lists the following symbols and

Adapted by permission of Hart Publishing Company, Inc., from its copyrighted volume VALUES CLARIFICATION: A Handbook of Practical Strategies for Teachers and Students by Sidney B. Simon, Leland W. Howe, and Howard Kirschenbaum. Also adapted from Sidney B. Simon, "Two Dozen Things I Love To Do," *Meeting Yourself Halfway*. Niles, Ill.: Argus Communications, 1974. Used with permission.

what they represent while simultaneously instructing the participants to:

1. Put a "$" by any item that costs three dollars or more each time you do it. Look for hidden costs. You may say it costs you nothing to take a walk in a state park, but getting to the park involves gas and car expenses.

2. Put a "10" by any item you would not have been doing ten years ago.

3. Put an "X" by any item you would like to let others know you do.

4. Put an "M" by each of the items that you have actually done in the last month.

5. Put a "T" by any of the items you spend at least four hours a week doing.

6. Put an "E" by any item you spend time reading about, thinking about, worrying about, or planning for.

7. Put a "C" by any of the items you consciously choose over other possible activities.

8. Put a "G" by the items that you think help you to grow as a person.

IV. The facilitator tells the participants that the more markings they have put next to an item, the more likely it is that the activity is a value for them. He adds that this list is not a compilation of their values but, rather, is an indication of where their values lie. He then instructs them to rerank their lists according to the number of marks next to each item—the activity with the most marks being first, etc. (Ten minutes or less.)

V. Participants are then instructed to compare their first ranking with their second ranking and to note:

1. Do your rankings match?

2. Is your highest value in your first ranking the one that has the most marks next to it?

3. Can you see any patterns in your list?

4. Have you discovered anything new about yourself as a result of this activity?

5. Is there anything you would like to change about your preferences as a result of this activity?

VI. The participants are divided into small groups (four to eight persons) to share what they have learned about their values. (Forty-five to sixty minutes.)

Variations

I. For a longer session, the list may be titled "Twenty Things I Love to Do."

II. For step III above, the symbols may be changed to represent different meanings, and the participants may be instructed to:

1. Put an "R" by any item that involves some risk. The risk may be physical, intellectual, or emotional.

2. Put a "P" by the items you do with other people and an "A" by the items you do alone.

3. Put a "PL" by any item that requires planning.

4. Put a "J" by any item that brings you joy or pleasure.

5. Put an "I" by any item that involves intimacy with another person.

6. Put a "U" by any item you enjoy doing but which most people do not do. (The "U" stands for your uniqueness.)

7. Put an "MT" by any item you would like to spend more time doing.

8. Put an "SA" by any item you think a self-actualized person would do.

9. Put a "B" by any item that you would like to become.

10. Put an "F" by any item you think will still be on your list five years from now.

11. Put an "L" by any item you would want to share with a loved one. Write the first name of the person with whom you actually do share that item.

Any additional criteria may be used, providing the participants' marks are positive indications of a value. (Positive indicators allow the marks to be added up for evaluation.)

III. The facilitator may introduce the value indicator "tools" by giving the following instructions:

1. List the top ten values you feel you hold in order of their priority.

2. List the ten "tools" you use the most. ("Tools" refers to any instrument, machine, or special piece of equipment, such as hiking shoes, car, tennis racket, typewriter, television, pencil, slide rule, etc.)

3. Rank the tools from first to tenth on the basis of the most used to the least used.

4. Furnish the following data about your tools:
   a. Put the price of the tool next to it.
   b. Put a "12" by any tool you would not have been using twelve years ago.
   c. Put an "X" by those tools about which you like to tell other people.
   d. Next to each tool, put the date of the last time you used it.
   e. Next to each tool, put a number that represents the hours you spend using the tool per week.
   f. Put an "E" by any tool you daydream about, read about, worry about, or plan for.
   g. Put a "G" by any tool you think helps you grow as a person in some way.

5. Now compare your ranking of the tools in step 3 with the markings generated from the information requested in steps 4a through 4g. Do they match? Compare the ranking of values in step 1 with the ranking of tools in step 3. The presupposition here is that since most values require some kind of tool to put them in action, you can tell something about your values by the tools you use most. In other words, if you value exercise, your most often-used tool may be hiking boots or a bicycle or a sweatsuit, etc.

6. Now form small groups and share what you have learned about your values.

WORKSHEET I: Love List

Make a list of ten things that you love to do in your life. In other words, of all the things you do in your life, list the ten you love to do most.

_____   1. _____   _____

_____   2. _____   _____

_____   3. _____   _____

_____   4. _____   _____

_____   5. _____   _____

_____   6. _____   _____

_____   7. _____   _____

_____   8. _____   _____

_____   9. _____   _____

_____   10. _____   _____

2. VALUE AUCTION

Goals

I. To determine those life values that are of greatest importance to participants.

II. To explore the degree of trust among participants.

III. To examine the phenomena of competition and cooperation.

IV. To invite consideration of how life values affect decisions concerning personal needs and aspirations.

Group Size

Twelve participants and one auctioneer.

Time Required

Forty-five minutes.

Materials

Variation 1:

I. A Value Auction Sheet for each participant.

II. A pencil for each participant.

Variation 2:

I. Ten tokens for each participant. These may be poker chips, paper clips, pennies, or whatever is handy.

II. A Value Auction Handout and a Value Auction Rule Sheet for each participant.

III. A pencil or pen for each participant.

IV. One small box (such as a shoe box) labeled "Token Bank."

V. Chalkboard and chalk.

Physical Setting

A large room with movable chairs.

This structured experience was originally designed by Roy W. Trueblood and Robert Rodgers. Used with permission.

Variation 1

Process

I. The facilitator (auctioneer) passes out a Value Auction Sheet to each participant and explains the goal of the activity. Each person is "given" $5,000 and is instructed to work independently and to use the first column to budget this amount for the listed items of value.

II. When budgeting is finished, the facilitator auctions off the items in random fashion or by asking the group to focus on items of value. The items should not be auctioned off in order.

III. Bids should be in increments of no less than $100. Participants are cautioned to keep track of their "bank balance." The use of column two, "Highest Amount I Bid," is important to help participants recall their interest in various items.

IV. When an item is sold, the highest bid is recorded by everyone in column three along with the initials of the person who bought it.

V. When all items are auctioned off, the facilitator processes the activity, focusing especially on the following questions:
1. Did you get what you wanted? If not, why not?
2. How did you feel about competing for what you wanted?
3. Did you spend all your money or do you have any left? How much? Why?
4. What did you learn about your personal value system?

Variation 2

Process

I. The facilitator (auctioneer) explains the goals of the activity; participants will be engaging in an exercise that invites them to consider those life values that are of greatest importance to them personally and the meaning those values have for them in the conduct of their everyday lives.

II. Each participant is given a pencil and a Value Auction Handout. The facilitator explains that this list is intended only as a partial listing of values, and he encourages participants to add any value not specifically listed that is important to them. Participants are then asked, on their own, to place a check mark by the three values of greatest importance to them and to make a note of any value they wish to add to the list. Persons may check more than three values if they wish.

III. As participants make their choices, the facilitator writes the values listed in the handout on the chalkboard.

IV. After the lists have been reviewed and checked, the facilitator asks for additions, indicating that this is the only time in the exercise when additions will be accepted. As additions are suggested, he adds them to the list on the chalkboard—unless they clearly overlap with a value already listed. Time is allotted for persons to talk about values of mutual interest.

V. The facilitator (auctioneer) then places the bank (containing enough tokens so that each participant can have ten) on the floor and asks that each person draw ten tokens from the bank, making it clear that all tokens have equal value.

VI. The Value Auction Rule Sheet is passed out as participants draw their tokens.

VII. After answering any questions, the facilitator (auctioneer) begins the auction by selling the first value to the highest bidder(s). The facilitator (auctioneer) then asks that the buyer(s) place the payment due in the bank. He writes the selling price and the buyer's(s') name on the chalkboard beside that value. Following the same procedure as above, always trying to obtain the highest selling price possible, the facilitator (auctioneer) sells the remaining values in the order listed on the chalkboard. He does not check to see how many tokens are actually deposited in the bank. (Allotted time for the auction should be about thirty minutes.)

VIII. The facilitator (auctioneer) then initiates a discussion focusing on the real meaning of the tokens; on what happened between and among participants during the auction; and on the meaning to participants of the values purchased. Participants are then given the task of deciding what they want to do with the values they have purchased.

VALUE AUCTION SHEET

	Amount I Budgeted	Highest Am't I Bid	Top Bid
1. A satisfying and fulfilling marriage	_____	_____	_____
2. Freedom to do what I want	_____	_____	_____
3. A chance to direct the destiny of a nation	_____	_____	_____
4. The love and admiration of friends	_____	_____	_____
5. Travel and tickets to any cultural or athletic event as often as I wish	_____	_____	_____
6. Complete self-confidence with a positive outlook on life	_____	_____	_____
7. A happy family relationship	_____	_____	_____
8. Recognition as the most attractive person in the world	_____	_____	_____
9. A long life free of illness	_____	_____	_____
10. A complete library for my private use	_____	_____	_____
11. A satisfying religious faith	_____	_____	_____
12. A month's vacation with nothing to do but enjoy myself	_____	_____	_____
13. Lifetime financial security	_____	_____	_____
14. A lovely home in a beautiful setting	_____	_____	_____
15. A world without prejudice	_____	_____	_____
16. A chance to eliminate sickness and poverty	_____	_____	_____
17. International fame and popularity	_____	_____	_____
18. An understanding of the meaning of life	_____	_____	_____
19. A world without graft, lying, or cheating	_____	_____	_____
20. Freedom within my work setting	_____	_____	_____
21. A really good love relationship	_____	_____	_____
22. Success in my chosen profession or vocation	_____	_____	_____

VALUE AUCTION RULE SHEET

During this Value Auction you will have the opportunity to use your ten tokens to buy, and thus to own, any of the values listed on the chalkboard—if your bid is highest. Owning a value means you have full rights and privileges to do with the value whatever you so choose at the conclusion of the exercise. Keep in mind the following rules:

1. There is no limit to the number of values that may be bought.

2. You may elect to pool your resources with other persons in order to purchase a particularly high-priced value. This means that two, three, four, or more persons may extend a bid for any one value.

3. The auctioneer's task is to collect the highest number of tokens possible in the course of the auction. After the auction has begun, no further questions will be answered by the auctioneer.

4. Only tokens will be accepted as payment for any value purchased.

VALUE AUCTION HANDOUT

Take a few minutes to think about the meaning to you of the items listed below and then decide on three or four which are of most importance to you personally. Indicate these items with a check (√) mark. At the bottom of the page add any value that is important to you but not specifically listed.

_____ Self-sufficiency

_____ Influencing others

_____ Exerting power over things (growing gardens, programming computers, fixing broken machines, etc.)

_____ Giving love

_____ Being spontaneous

_____ Having an active and satisfying athletic life

_____ Opportunities for adventure

_____ Having an active and satisfying sex life

_____ Good health

_____ Large family

_____ Wealth

_____ Approval by the opposite sex

_____ Intellectual stimulation

_____ Keeping physically attractive

_____ Contentment with doing nothing for long periods of time

_____ Prestige

_____ Maintaining long-term friendships

_____ Receiving love

_____ Having a close and supportive family life

_____ Making time for hobbies

_____ Indulging in frequent travel

_____ Freedom on the job to come and go as one pleases

_____ Spiritual fulfillment

_____

_____

3. VALUE VOTING

Goals

I. To give participants an opportunity to publicly affirm their stand on a value issue.

II. To quickly surface those value issues about which the group is most concerned.

III. To give participants the chance to know whether others have similar value concerns.

IV. To explore and understand one's own values as well as the values of others.

Group Size

Thirty to forty people.

Time Required

Approximately one and one-half to two hours.

Physical Setting

A large room with movable chairs.

Process

I. The facilitator selects a value issue on which the group is divided. He instructs participants to form two circles: one consisting of those who respond positively to the value and the other consisting of those who respond negatively to the value. Participants are given twenty minutes to explore why they are in the group they have chosen. They may change from one circle to the other.

II. (A) Participants form a fishbowl arrangement, with the "yes" group on the inside. "Yes" group members take twenty minutes to discuss their choosing, acting, and prizing of the value in question.

(B) The "no" group enters the fishbowl and takes twenty minutes to discuss their reasons for saying "no" to the value. The "yes" group observes this discussion.

III. The two groups meet separately to discuss what they have heard from the other group and to elect three representatives who will discuss the value issue with three representatives of the other group. (Twenty minutes.)

IV. The six representatives meet to discuss the value in question. Their goal is to clarify the alternatives, consequences, acting on, and prizing of the value. All other participants observe this discussion. (Twenty minutes.)

Variations

I. Psychic Power Discovery. This is a simulation discussion that employs value voting.[1]

1. The following instructions are read to the group:

In 1984, a group of scientists discovered a method of teaching people how to develop their psychic powers so that anyone could know any other person's deepest thoughts and feelings that he is consciously experiencing. The scientists presented a proposal to the United Nations that all the countries of the world make use of this discovery by teaching their citizens how to develop this power. The power could be used only if all the countries accepted the proposal so that no country would have an advantage. Each nation attempted to assess the impact of the use of this discovery on all aspects of personal, national, and international life in order to vote correctly on whether this proposal should be accepted. They investigated and explored the impact of this proposal on:

a. a person's sense of self-identity and self-protection;

b. personal interactions: friends, companions, enemies;

c. marriages;

d. intergroup and interracial relationships; and

e. national and world politics.

If you were part of a U.S. commission studying this question, what would your predictions be about the possible good and bad consequences of the use of this psychic power by all the peoples of the world? As a member of this commission, would you vote for the acceptance and use of this psychic power by all the nations of the world? Try to make some concrete predictions of what would happen by the use of this psychic power in all the five areas of life indicated above.

2. Process:

a. The participants break into dyads and discuss their predictions about the use of this power in the five areas of life. A

---

[1]Submitted by Francis Elmo.

decision for or against the use of this power is made. (Ten minutes.)

b. The total group is re-formed and takes a vote of those voting for and against the proposal. The "yes" group and the "no" group meet separately and caucus on their reasons for taking their particular position. (Ten minutes.)

c. Each side returns to present its position to the other side. First the "yes" group speaks and the "no" group listens; then the "no" group speaks and the "yes" group listens. (Ten minutes for each side.)

d. Each side reverses positions and presents the arguments of the other side. (Ten minutes for each side.)

e. Re-voting and general discussion. Each member has an opportunity to re-vote on the proposal. General discussion follows on what the experience reveals about individuals' fears of self-revelation and openness. (Twenty minutes.)

II. As an introductory exercise, the group may brainstorm a list of values that the members are interested in dealing with during the workshop or series of sessions. Members may simply take a vote on each value presented for consideration or may list the values and then rank them.

III. If a group is discussing a value issue and there are only a few members who are vocal about their position on the issue, it is helpful to the facilitator and the group to find out where the rest of the members stand. The facilitator may call for a value vote to show who says "yes," who says "no," and who has "no opinion" on the value issue. The vote may be dramatized by having people physically move into corners ("yes" corner, "no" corner, "no opinion" corner) to indicate their votes.

If most of the participants vote yes or no, the discussion is continued or an activity is introduced to further the discussion. If most of the participants hold no opinion on the issue, the facilitator moves on to another value issue. If the votes (or groups in each corner) are split fairly evenly, participants are given twenty minutes to clarify their positions. Then three representatives from each position meet in the center of a group-on-group arrangement to discuss the value.

IV. A facilitator who is working with an intact group can devise a list of value issues that are inherent in the group and that will tap the group's interest. For example, in dealing with a group of managers who have expressed an interest in the value of openness, the facilitator may present Worksheet I and have members vote on the items by going to a particular spot in the room. A few minutes is spent discussing the responses; then members record their responses by marking the worksheets, which they attach to the front of their clothing with tape. Participants mill about the room reading and discussing one another's worksheets.

For a discussion of value voting and a description of similar techniques, see L. E. Raths, M. Harmin, and S. B. Simon, *Values and Teaching*, Columbus, Ohio: Charles E. Merrill, 1966, pp. 152-154.

References

Pfeiffer, J. W., & Jones, J. E. (Eds.). *The 1972 annual handbook for group facilitators*. La Jolla, Calif.: University Associates, 1972, pp. 75-85.

Simon, S. B., Howe, L. W., & Kirschenbaum, H. *Values clarification*. New York: Hart Publishing, 1972, pp. 38-57.

WORKSHEET I: Manager Intervention Style

1. If one of my assistants criticized a decision of mine, I would:
   a. not talk to him but wait until he talks about it.
   b. tell him my reasons for the decision.
   c. let him talk while I listen.
   d. talk out a position both of us can accept.
   e. try to persuade him to agree with my viewpoint.

2. If a group of employees wanted to talk about a new policy that they were upset about and disagreed with, I would:
   a. just listen to them and let them get it off their chests.
   b. explain to them the reasons for the policy.
   c. help them to talk about their concerns.
   d. tell them they will just have to adjust to the new situation.
   e. be friendly with them but win them over to my position.

3. If I discovered that I disliked a new assistant with whom I had to work, I would:
   a. talk to others to see if they agree with me, then talk to him.
   b. express my feelings to the new assistant first.
   c. invite him to talk about our relationship.
   d. try to understand him and overcome my negative feelings.
   e. avoid him but be polite when I have to work with him.

4. If two of my assistants often got into heated arguments, I would:
   a. stay out of it.
   b. try to smooth over their feelings and keep them working.
   c. present to them my views on the disagreement.
   d. tell them to work out an equitable solution.
   e. try to help them explore their differences and come to an understanding with each other.

4. VALUE RANKING

Goals

I. To practice choosing from alternatives after considering the consequences.

II. To explore alternatives and consequences with others.

III. To publicly affirm choices.

IV. To discover that many issues demand more consideration than a person usually gives.

Group Size

Group size is optional. Members work together and then form subgroups of six to eight.

Time Required

Flexible. Fifteen minutes is needed for the basic activity, and forty-five to sixty minutes should be allotted for small-group sharing.

Materials

I. Pencils and paper for each participant.

II. Worksheets (as required) or newsprint and felt-tipped pens.

Physical Setting

A large room with movable chairs.

Process

I. The facilitator discusses value ranking and how it helps a person to order his values according to their priority, thereby enabling him to discover which values are most important to him. He explains that "forced ranking" means that *only one item* may be ranked in each category: number one (most important), number two, and so on.

II. The facilitator presents the group with the items to be ranked, using one of the following methods:

1. He may read the open-ended statements out loud, allowing time between each statement for the participants to complete the blank sentence and then to rank the items from one to

Adapted by permission of Hart Publishing Company, Inc., from its copyrighted volume VALUES CLARIFICATION: A Handbook of Practical Strategies for Teachers and Students by Sidney B. Simon, Leland W. Howe, and Howard Kirschenbaum.

four. This method allows time for the participants to reflect on their choices.

2. He may list each group of statements on newsprint, allowing time for participants to complete all the statements and rank them before listing the next group. This method avoids the necessity of repeating the basic statement and avoids flooding the group with too many items at one time.

3. He may make copies of the open-ended statements so that they may be used as a worksheet.

The items to be ranked are:

1. I would like to be seen as a person who is strong.
   I would like to be seen as a person who is warm.
   I would like to be seen as a person who is intelligent.
   I would like to be seen as a person who is _____.

2. I would like to be able to express anger.
   I would like to be able to express love.
   I would like to be able to express joy.
   I would like to be able to express _____.

3. My highest value is success in my work.
   My highest value is my relationship with my spouse.
   My highest value is creating a loving family.
   My highest value is _____.

4. I spend most of my free time alone.
   I spend most of my free time with my family.
   I spend most of my free time doing more work.
   I spend most of my free time _____.

5. The greatest fear I have is of failure.
   The greatest fear I have is of being rejected by others.
   The greatest fear I have is of becoming useless.
   The greatest fear I have is of _____.

III. Participants form subgroups of six to eight persons each and share their rankings for forty-five to sixty minutes. (If time is short, this step may be abbreviated.)

IV. Each individual takes approximately fifteen minutes to formulate a plan to work on a value issue that has emerged.

 V. Participants return to their subgroups to share plans and explore alternatives and consequences, using each other as resource persons. (About one hour.)

Variations

I. Ranking is useful as an introduction to a series of sessions or as the basis for a longer workshop. In a small group of up to fifteen persons, each member lists his ten most important values on newsprint, force ranks them, and hangs his sheet on the wall. A larger group is divided into subgroups of six to eight which are then instructed to produce a group consensus ranking. The group members take thirty to forty-five minutes to share what they think are their most important values and to define or describe these values. The facilitator then takes the top values and, working downward, designs each of the succeeding sessions to deal with one of the values identified by the group.

II. (A) The facilitator passes out Worksheets I-A and I-B and instructs participants to add four more topics of their own choice to complete numbers 17 through 20 on Worksheet I-A. He allows time for completion of this task.

(B) Participants are told to force rank the twenty questions on the basis of how strongly they feel about each topic and then to place the underlined word from each sentence in the appropriate box on Worksheet I-B.

III. (A) The facilitator passes out Worksheet II, Priority Rankings, and instructs participants to rank the items in each category according to their priority from 1 to 4. The group members can be divided into subgroups of six to eight members each to discuss their rankings.

(B) Subgroups can determine priority rankings by consensus.

IV. (A) With a group that is familiar with the value-ranking technique, the facilitator may have the participants design their own instrument. This would consist of about six issues with the group suggesting three possibilities for each issue and leaving the fourth possibility open so that individuals can add their own.

(B) Participants then force rank the items and form small groups to share their rankings.

(C) Individual members then take time alone to plan how they might grow in one of the values that has emerged for them.

(D) Participants return to the small groups to share their plans and to use one another as resource persons to explore alternatives, consequences, and possible actions.

For similar techniques, see:

Hall, B. P. *Value clarification as learning process: A guidebook*, New York: Paulist Press, 1973, pp. 79-93.

Penney's Forum, Spring/Summer 1972, p. 24.

Simon, S. B., Howe, L. W., & Kirschenbaum, H. *Values clarification*. New York: Hart Publishing, 1972, pp. 58-93, 98-111.

Simon, S. B. *Meeting yourself halfway*. Niles, Ill.: Argus Communications, 1974, pp. 44-53.

WORKSHEET I-A: Value Statements

Rank the following statements from 1 to 20 according to how strongly you feel about them. The underlined word is to be written in the appropriate box on Worksheet I-B. Only one item may be placed in each box.

How do you feel about:

1. a *friend* who betrays a confidence after you have had a disagreement?
2. a family who will not take in an *elderly* parent unless the parent pays $30 a week toward expenses?
3. a man or woman who *remarries* one month after his/her spouse dies?
4. paying one to two percent more in taxes if it were to go toward reducing *pollution*?
5. *violence* on television during child-watching hours?
6. your son or daughter deciding to enter into a *nonconventional* living arrangement?
7. someone who could afford to but does not pledge support to the *religious* organization to which he belongs?
8. a family that moves to the *suburbs* to provide a better environment for the children, causing the father to be away from the family from three to four additional hours a day because of the time spent commuting?
9. someone who spends $4,000 on the *funeral* of a loved one?
10. a *court* decision ordering a woman who has been the principal provider in a family to pay alimony to her ex-husband?
11. an *employer* who advertises equal opportunities and equal salaries for women and then asks a young woman being interviewed if she is planning to be married soon?
12. a person who turns in a neighbor for *cheating* on his income tax?
13. having a new *jet port* constructed one mile from your home?
14. controlling population by *limiting* families to two children?
15. an *employee* who takes the day off but calls his office to say he is sick?
16. someone accepting a job that does not require as much *responsibility* as he is capable of, so that he will have free time to pursue outside activities?

Add other value issues about which you feel strongly and underline a key word.

    17. How do you feel about . . .

    18. How do you feel about . . .

    19. How do you feel about . . .

    20. How do you feel about . . .

WORKSHEET I-B: Value Boxes

Very Strong Feeling	Strong Feeling	Mild Feeling	Little Feeling
1	6	11	16
2	7	12	17
3	8	13	18
4	9	14	19
5	10	15	20

Adapted from *Penney's Forum*, Spring/Summer 1972, p. 24.

WORKSHEET II: Priority Rankings

1. For recreation, I prefer:
   _____ conversation in a small group.
   _____ doing something together in a group.
   _____ watching TV.
   _____ reading the newspaper.
   _____ add another: _____

2. In my work, I like to:
   _____ work by myself.
   _____ work with others.
   _____ delegate responsibility to others.
   _____ add another: _____

3. I like to use my free time to:
   _____ be by myself.
   _____ visit friends.
   _____ catch up on work.
   _____ add another _____

4. For my retirement I would want to:
   _____ work on my hobbies.
   _____ travel.
   _____ take a part-time job.
   _____ add another: _____

5. For the future I should:
   _____ stay just as I am.
   _____ take on new interests.
   _____ renew present interests.
   _____ drop some of my present interests.
   _____ add another: _____

6. With colleagues, it is best to:
   _____ keep quiet about yourself and your work.
   _____ ask for help, advice, or consultation when you need it.
   _____ be friendly but not talk about personal or important business matters.
   _____ tell them about yourself and your work.
   _____ add another: _____

7. I think that the best boss is one who:

_____ tells you what should be done.

_____ consults with you on important issues.

_____ persuades you to live up to your ideals.

_____ facilitates a consensus on important issues.

_____ add another: _____

8. Does this list leave out any important areas? If so, add other areas below, including possible approaches to that area, and then rank them as above.

5. VALUE CONTINUUM

Goals

I. To present or brainstorm the full range of alternatives in a controversial issue to be considered.

II. To clarify the many positions that have been taken on a controversial issue before polarization occurs.

III. To avoid oversimplified and "either/or" thinking by demonstrating the complexity of a situation.

IV. To increase freedom of choice by expanding the participants' awareness of the possible variables within an issue.

Group Size

Group size is optional. Members work together or as individuals and then form subgroups of six to eight to share.

Time Required

Tailored to the group and the purpose of the activity.

Materials

I. Pencils and worksheets (as required) for all participants.

II. Newsprint and felt-tipped pens.

Process

I. The facilitator explains the technique of the value continuum to the group and hands out copies of Worksheet I, II, or III.

II. The facilitator may solicit from the group members as many additional polar possibilities as they can think of. He lists these items on newsprint, making sure that each word is defined as it is listed, while participants fill in the blank spaces on their worksheets.

III. Each participant marks the value continuum on his worksheet, indicating where he stands in relation to each set of opposing items by making an "X" somewhere along the continuum line.

IV. The group then divides into subgroups of six to eight to share their value positions. (Thirty-five to forty minutes.)

Submitted by Lorraine Weber.

Variations

I. A continuum may be used to break down "either/or" thinking. It takes a problem that usually splits people into two oversimplified camps and demonstrates a wide range of intellectually defensible positions.

Preposterous positions can be set up at either end of the continuum. The hope is that these positions will be so extreme that no one would dare support them.

A value continuum on racial conflict is a good example. This exercise opens up a highly charged topic so it can be discussed with some degree of rationality. At one extreme is Super Separatist Sam, whose solution to the race problem is simply to ship every human being back to his original country. He advocates dismembering people whose ancestors came from two different countries. On the other end is Multi-Mixing Mike who would insist that all babies be distributed not to their original parents but to couples not of the baby's race. In addition, no couples of the same race can marry, and couples of the same race already married must be divorced and marry outside their race.

Between these extremes, participants must come face-to-face with where they stand on issues of integration, black power, de facto segregation, and so on.

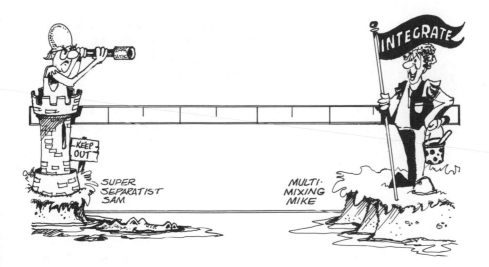

The group could also act out the value continuum physically, literally "taking a stand" on where they are.

II. (A) The value continuum may be used to present a group with the range of variables on an issue under study. In lieu of a lecture on the subject of dealing with feedback, for example, Worksheet I may be used to involve the participants in a study of the criteria for effective feedback. The blank spaces allow the group members to insert polar items that they think should be included in the continuum.

(B) In a human relations laboratory, the facilitator may extend the activity by directing the group members to practice giving feedback to one another.

III. (A) Worksheet II deals with interpersonal relationships. Group members may add additional items to the list.

(B) Once the members' positions on the continuum are filled in, subgroups of eight to ten are formed for members to share the experience.

IV. Worksheet III is an example of an actual group activity. The group created its own continuum and then made plans to act on it. The following format was used:

1. For twenty to thirty minutes, the group brainstormed a list of qualities of a good friend, placing "good" qualities on the left and their opposite characteristics on the right.

2. Each participant marked the continuum according to where he stood in terms of the qualities listed.

3. In subgroups, group members shared their markings and their interpretations of what the markings imply about them.

4. Each participant then took about fifteen minutes to select two or three qualities in which he was weak and to design a plan for improvement in those areas.

5. The subgroups re-formed and members used one another as resource persons to suggest alternatives, explore consequences, and suggest actions. This final step took almost an hour.

For a discussion of value continuum, see L. E. Raths, M. Harmin, and S. B. Simon, *Values and Teaching*, Columbus, Ohio: Charles E. Merrill, 1966, pp. 129-130, and B. P. Hall and M. Smith, *Value Clarification as Learning Process: A Handbook*. New York: Paulist Press, 1973, p. 131.

References

Hall, B. P. *Value clarification as learning process: A guidebook.* New York: Paulist Press, 1973, pp. 47, 129-131, 148, 159-165, and 239.

Simon, S. B., Howe, L. W., & Kirschenbaum, H. *Values clarification.* New York: Hart Publishing, 1972, pp. 116-126.

Weber, Lorraine, "How we teach values," *Colloquy*, 1970, *3* (1), pp. 8-12.

WORKSHEET I: Criteria of Effective Feedback

Descriptive	1	2	3	4	5	4	3	2	1	Evaluative
Specific	1	2	3	4	5	4	3	2	1	General
Sensitive	1	2	3	4	5	4	3	2	1	Insensitive
Inconsistent	1	2	3	4	5	4	3	2	1	Characteristic
Solicited	1	2	3	4	5	4	3	2	1	Imposed
Immediate	1	2	3	4	5	4	3	2	1	Delayed
Verified	1	2	3	4	5	4	3	2	1	Presumed
_____	1	2	3	4	5	4	3	2	1	_____
_____	1	2	3	4	5	4	3	2	1	_____
_____	1	2	3	4	5	4	3	2	1	_____

WORKSHEET II: Relationships With Others

Like to do things alone	1 2 3 4 5 4 3 2 1	Like to do things with others
Like to be a follower	1 2 3 4 5 4 3 2 1	Like to be a leader
Like to keep distant	1 2 3 4 5 4 3 2 1	Like to be close
Do not like to touch	1 2 3 4 5 4 3 2 1	Like to be touched
Do hobbies by self	1 2 3 4 5 4 3 2 1	Join clubs
Accept others' decisions	1 2 3 4 5 4 3 2 1	Uphold own decisions
Tend to be cold	1 2 3 4 5 4 3 2 1	Tend to be warm
Do not like to be bothered	1 2 3 4 5 4 3 2 1	Like to be asked to join
Like to make decisions	1 2 3 4 5 4 3 2 1	Dislike decision making
Need only self	1 2 3 4 5 4 3 2 1	Need others
Prefer to be told	1 2 3 4 5 4 3 2 1	Prefer to tell
_____	1 2 3 4 5 4 3 2 1	_____
_____	1 2 3 4 5 4 3 2 1	_____
_____	1 2 3 4 5 4 3 2 1	_____

WORKSHEET III: An Example From an Actual Group: Friendship

Warm	1 2 3 4 5 6 7 8 9 10 9 8 7 6 5 4 3 2 1	Undemonstrative
Generous	1 2 3 4 5 6 7 8 9 10 9 8 7 6 5 4 3 2 1	Selfish
Caring	1 2 3 4 5 6 7 8 9 10 9 8 7 6 5 4 3 2 1	Indifferent
Acceptant	1 2 3 4 5 6 7 8 9 10 9 8 7 6 5 4 3 2 1	Intolerant
Honest	1 2 3 4 5 6 7 8 9 10 9 8 7 6 5 4 3 2 1	Deceitful
Open	1 2 3 4 5 6 7 8 9 10 9 8 7 6 5 4 3 2 1	Secretive
Loyal	1 2 3 4 5 6 7 8 9 10 9 8 7 6 5 4 3 2 1	Disloyal
Considerate	1 2 3 4 5 6 7 8 9 10 9 8 7 6 5 4 3 2 1	Thoughtless
Understanding	1 2 3 4 5 6 7 8 9 10 9 8 7 6 5 4 3 2 1	Unsympathetic
Patient	1 2 3 4 5 6 7 8 9 10 9 8 7 6 5 4 3 2 1	Impatient
Trusting	1 2 3 4 5 6 7 8 9 10 9 8 7 6 5 4 3 2 1	Suspicious
Courteous	1 2 3 4 5 6 7 8 9 10 9 8 7 6 5 4 3 2 1	Disrespectful

6. INCOMPLETE VALUE SENTENCES

Goals

I. To stimulate participants to begin to think about what they believe in and where their values lie.

II. To enable participants to share their value indicators with others.

III. To raise the participants' interest in further exploration of their values.

Group Size

Any number of participants.

Time Required

Three minutes per sentence; about one and one-half hours for the entire activity.

Materials

Pencil and paper for each participant.

Process

I. The facilitator invites the participants to complete the following sentences, which he reads aloud. As he reads, the facilitator pauses between each sentence to allow time for participants to write down their responses.

1. The greatest joy in my life is . . .
2. The biggest decision I ever made was . . .
3. My constant worry is . . .
4. As a child I dreamed of . . .
5. The thing I love most about life is . . .
6. What I would like to change in my life is . . .
7. The three things in the world I would change are . . .

II. The facilitator divides the participants into subgroups of six to eight and instructs them to share their responses, but only to the extent that they feel comfortable in doing so.

III. The facilitator then helps the group to explore the values underlying their responses. He may have a volunteer read the re-

Adapted by permission of Hart Publishing Company, Inc., from its copyrighted volume VALUES CLARIFICATION: A Handbook of Practical Strategies for Teachers and Students by Sidney B. Simon, Leland W. Howe, and Howard Kirschenbaum.

sponses, and then he may model how to perceive the values implicit in a statement. The group will need twenty to thirty minutes to discuss the value implications.

Variations

I. Numerous other open-ended statements may be used in this technique. Some examples are:

1. My hero is . . .
2. I hope someday to . . .
3. I spend most of my time . . .
4. The greatest sorrow of my life is . . .
5. I would prefer to be . . .
6. A friend is someone who . . .
7. My father taught me to . . .
8. My mother always said . . .
9. If in the future I were physically disabled I would . . .
10. My favorite kind of person is . . .
11. People are usually . . .
12. Money is the most . . .
13. What I would like to do most is . . .
14. My family likes to . . .
15. To me, kids are . . .
16. The work I do is . . .
17. The people I work with are . . .
18. If I were the boss I would . . .
19. When I retire I am . . .
20. My favorite place in the world is . . .
21. The person who influenced me the most taught me to . . .
22. I daydream the most about . . .
23. I want to die when I am . . .
24. My greatest accomplishment in life is . . .
25. The one quality I would most like to develop is . . .

II. (A) When a longer list of sentences is used, the facilitator may list them on a worksheet. After participants have completed the sentences, they are asked to respond to the following:

1. Put a "P" in front of those statements about which you arc most proud.

2. Put a "PA" in front of those statements that you would publicly affirm.

3. Put a "CA" in front of those statements for which you have considered alternatives.

4. Put a "TC" in front of those statements for which you have thoughtfully considered the consequences.

5. Put a "CF" in front of those statements you know for sure you have chosen freely; that is, values that were not imposed on you by your parents or by anyone else or by any circumstances.

6. Put an "A" in front of the statements you have actually acted on.

7. Put a "PB" in front of those statements that indicate a pattern of behavior, a habit.

8. Put a "G" in front of those statements that you know help you to grow as a person in some way.

(B) The participants form small subgroups to share their responses and evaluations.

7. COAT OF ARMS

Goals

I. To use such value indicators as aspirations, interests, hopes, beliefs, and activities to discover the values of the individual and/or of the group as a whole.

II. To use fantasy in discovering one's values.

III. To provide an opportunity for the individual to publicly affirm some important values in his life.

IV. To use the value-clarification consensus or comparison activity as a team-building intervention with an intact group.

Group Size

Unlimited number in groups (preferably intact units) of approximately eight persons each.

Time Required

Approximately one and one-half hours.

Materials

I. Paper and a pencil for each participant. (Crayons are optional.)

II. A General Instruction Sheet for each participant.

Process

I. The facilitator states that this is a value-clarification activity. He explains that, historically, a coat of arms was a symbol of what a family stood for (valued) and that pictures were usually used to depict these virtues or achievements; e.g., a lion meant strength, a cross meant defense of the church, etc.

II. Each participant receives a General Instruction Sheet. The facilitator points out that the coat of arms outline has six sections—each of which will be used to depict the response to a specific statement. He assures them that artistic ability is not important and that stick figures may be used. He says there will be five minutes allotted for each section and that pictures or symbols, not words, are to be used unless otherwise indicated. (Participants may prefer to draw their own coat of arms outlines.

Adapted by permission of Hart Publishing Company, Inc., from its copyrighted volume VALUES CLARIFICATION: A Handbook of Practical Strategies for Teachers and Students by Sidney B. Simon, Leland W. Howe, and Howard Kirschenbaum.

Likewise, the times may be adjusted, depending on the needs and experience of the particular individual or group.)

III. (A) Allowing five minutes for the completion of each task, the facilitator directs the participants to draw, within the designated area of their coat of arms, a picture or symbol of:

Area 1—Your favorite activity.

Area 2—Your highest achievement last year.

Area 3—Your saddest experience last year.

Area 4—Something you would be willing to die for.

Area 5—What you would do with the last year of your life if you had only one year to live and had all the money and power necessary to do whatever you wished.

For Area 6, each participant is directed to write three words that describe him *as a person*. The facilitator adds that these may be the three words each participant would like as his own epitaph.

(B) In some cases, the facilitator may want to utilize alternative statements. The following are suggested:

(To be depicted by a picture or symbol)

1. Your greatest personal achievement to date.

2. The one thing that other people can do to make you happy.

3. Something you are striving to become (or for).

4. What you would want to do if you knew you would be successful at it.

(To be written)

1. The three things you would most like to have said about you (if you died today).

2. The personal motto by which you (try to) live.

(Step III will take about one-half hour.)

IV. (A) If this activity is used as an introductory experience, the participants may be divided into groups of eight to ten persons and asked to pin their coats of arms to the front of their clothing. Subgroup members ask one another *succinct* questions about symbols they do not understand and then team up with others whose depicted values most interest them.

(B) The small groups share what they have discovered about their values. (Forty-five to sixty minutes.)

V. The facilitator may lead a brief discussion on value indicators.

VI. (A) If the experience is an introductory one, it may be closed with a brief discussion of what happened. If it is part of a workshop or series of experiences, participants may select one value that has emerged and check it against the criteria for a value. (Fifteen to twenty minutes.)

(B) Participants then return to their groups of eight and use one another as resource persons to explore alternatives, consequences, and actions of the values they have surfaced during the session.

Variations

I. (A) To adjust the experience for an intact work or family group, the facilitator begins by stating the importance of group members' knowing one another's values and reaching agreement on common goals and values.

(B) Large groups are divided into subgroups of six to eight persons each.

(C) Each of the intact group or family members creates an emblem, symbolizing his or her purposes, values, characteristics, etc.

(D) When subgroup members have completed their own emblems, they join to share their answers and create a group consensus emblem, using newsprint and felt-tipped pens. The time required for this activity will vary according to the number of participants; generally it will take one-half hour for the completion of personal emblems and one to one and one-half hours for the group drawing. Groups are instructed to avoid averaging, majority voting, or trading off, and to strive for real consensus.

(E) Each group then displays its insignia, and all participants discuss what they learned about one another's values and how they feel about their group.

II. Family groups may be given an additional fifteen to twenty-five minutes to discuss the common or discrepant values that emerge.

III. With married couples, each person draws a personal coat of arms and one for his or her spouse. They then share these insignias. They may then draw a mutual coat of arms.

Note: Many items in the personal value-clarification inventory may be successfully modified for intact groups and families. In addition, the following suggestions may also be incorporated into a coat of arms for such groups.

1. The most fun you have had with your colleagues/family in the past_____ months.
2. What keeps you from enjoying your group/family.
3. The most uncomfortable time you have had with your group/family in the past_____ months.
4. Something that is common to all or most of your colleagues/family members.
5. Something you and some of your colleagues/family members often do together.
6. The changes you would like to see in your group/family.
7. What your group/family does best together.

Reference

Simon, S. B. *Meeting yourself halfway*. Niles, Ill.: Argus Communications, 1974, pp. 81-83.

GENERAL INSTRUCTION SHEET

Because this is an experience in value clarification, it is important that you be as candid as possible in your responses to the statements.

Do not worry about your artistic ability—it is what you are depicting that matters.

You will have five minutes to draw a picture or symbol in each section of your coat of arms.

8. VALUE THOUGHT SHEET

Goals

I. To discover current concerns and meaningful events and the values that may be emerging from them.

II. To consult with others about possible alternatives to and consequences of these concerns and events.

III. To encourage the presentation of value topics by the participants, rather than by the facilitator.

IV. To share honest thoughts about values with other people.

Group Size

Any size group may be used; however, the activity works especially well with groups of eight to ten people.

Time Required

Approximately two and one-half hours, depending on the size of the group.

Materials

Paper and pencils for all participants.

Process

I. The facilitator directs participants to take twenty minutes or one-half hour to reflect on what has happened to them in the past week. He asks them to choose from one to three significant thoughts or events that have arisen during the week and write about them.

II. Subgroups of six to eight persons are formed so that members can share their writings with one another. This sharing should take about one-half hour.

III. The facilitator gives a ten-minute lecturette on the criteria for a value and value indicators, emphasizing the development of values in a person's life.

IV. In light of the lecturette, participants review what they have written to discover the values that have emerged or are emerging for them. Each person chooses one value in which he would like to grow and makes a tentative plan that he would be willing to share with the group. (Fifteen to twenty-five minutes.)

V. Participants return to their subgroups and use one another as resource persons to explore their emerging values, possible alternatives, consequences, actions, and plans for implementing these values.

Variation

(A) Any number of questions may be asked in steps IV and V to structure the activity for a particular group. Some of these questions are:

1. Is the emerging value a cherished belief or attitude of yours?
2. Do you like the value that has emerged; did you expect this value?
3. During the past week, did any pattern emerge in which the same value surfaced in different circumstances?
4. Would you be willing to write a letter to the editor of the local newspaper concerning this value?
5. What value did you expect to emerge that did not? Is that value indicator really acted on in your life? What can you do to make it a value?
6. As you review your thought sheet, what general outlooks on life present themselves?
7. Does your thought sheet reflect and agree with your stated life goals?
8. Does your thought sheet reflect what you say you believe in?
9. Does your thought sheet reveal a problem area for you?
10. What is the probability of the value that emerged in this thought sheet becoming a recurring theme in your life?
11. Does this thought sheet accurately represent the way you spend your time, thus reflecting your more important values?
12. If you have thought frequently about the prominent idea or value in this thought sheet, have you done anything to act on it?

(B) Other techniques may be employed in steps IV and V, depending on the needs of the group and the purpose (problem solving, team building, etc.) of the activity.

For a discussion of value thought sheets and a description of similar techniques, see L. E. Raths, M. Harmin, & S. B. Simon, *Values and teaching*. Columbus, Ohio: Charles E. Merrill, 1966, pp. 130-134.

Reference

Simon, S. B. *Meeting yourself halfway*. Niles, Ill.: Argus Communications, 1974, pp. 13-16.

9. VALUE AUTOBIOGRAPHY

Goals

I. To explore the values a person has formed.

II. To practice using the value indicators as a means for discovering those values.

III. To share values with others in order to increase personal growth and develop group trust and cohesiveness.

Group Structure

Participants work independently, then form groups of four to six.

Time Required

Approximately three hours, depending on group size.

Materials

A pencil and a notebook for each participant.

Process

I. The facilitator invites the participants to reflect on their lives. (He does not talk about values or value indicators at this point.) He tells the participants that they are to take one-half hour to make three graphs, as follows:

1. A graph of family life.

2. A graph of work life (including education and occupations).

3. A graph of personal use of time (which may include hobbies, sports, reading, friends, etc.).

II. Participants choose other persons with whom they would feel comfortable in sharing some of their personal history. Each person takes about five minutes "air time"; a group of four will need twenty minutes, etc.

III. Participants resume individual work. The facilitator directs them to reflect on the lives they have described by asking themselves the following questions and listing the answers chronologically:

1. What major decisions have I made in my family life, work life, and personal life?

2. What goals, aspirations, attitudes, interests, feelings, beliefs,

61

worries, and activities does my history in family life, work life, and personal life reflect?

3. Where or on what have I spent most of my time, money, and energy?

This step takes about one-half hour.

IV. The facilitator gives a brief lecturette on discovering values through value indicators.

V. Participants return to their small groups and use one another as resource persons to explore and discover the values they have been attempting to live out in their lives. Each participant is given at least fifteen minutes air time; a group of four would need about an hour, etc.

VI. The facilitator may call for a general sharing session and invite the participants to share the most important or most interesting discovery they have made about themselves and their values. Participants are reminded that they need share only what they feel comfortable sharing.

Variations

I. The facilitator may allow the participants one-half hour to write a simple autobiography in terms of their family life, work life, and personal life, and then introduce the Value Indicator Search activity (p. 133) for about one-half hour. Participants then form groups of four to six to share (for one-half hour) what they have discovered.

II. (A) The autobiography may be structured in many ways. The following directions and questions may be used as one set of guidelines. (Note: Some of the following points may seem repetitive, but they are meant to point out different aspects of similar issues so that participants may choose the ones that most closely represent their experiences.)

1. What are some major decisions you have made that have had happy consequences for you?

2. What is the ideal version of what you are going to do with your life?

3. What would you be willing to die for?

4. List the kinds of activities on which you have spent a considerable amount of time.

5. What have you spent most of your money on?

6. If you could be doing anything you wanted, what would you be doing?

7. What is your favorite activity?

8. Give the history of your favorite hobby or leisure-time activity.

9. What has been your plan of life?

10. What goals do you set for your life?

11. List and describe the important people in your life.

12. Describe your philosophy of life.

13. List things, ideas, and people you like and describe what you like about them.

14. Of the values that you received from your family, which have you accepted and which have you rejected?

15. What do you consider to be your potential in life?

16. What kind of person will you be ten years from now?

17. What would be an ideal vacation for you?

(B) The facilitator may want to reproduce these questions and hand them out to participants, or he may prefer to list them ahead of time on newsprint.

III. If an autobiography activity is used during a two-day or three-day workshop, participants may write on paper what they have discovered about their values, pin the paper on their clothing, and mill around silently reading one another's discoveries. This generally serves to increase trust, understanding, and cohesiveness among group members.

IV. Other activities may be used in combination with the Value Autobiography. The Life Planning activity (Pfeiffer & Jones, 1974) and Coat of Arms (p. 53) are examples.

For similar techniques, see:

Pfeiffer, J. W., & Jones, J. E. (Eds.). *The 1974 annual handbook for group facilitators.* La Jolla, Calif.: University Associates, 1972, pp. 101-115.

Progoff, I. *At a journal workshop.* New York: Dialogue House Library, 1975.

Raths, L. E., Harmin, M., & Simon, S. B. *Values and teaching.* Columbus, Ohio: Charles E. Merrill, 1966, pp. 140-141.

10. VALUE-CLARIFICATION JOURNAL

Goals

I. To create an awareness of the process of one's value development.

II. To crystallize one's learning from a series of value experiences or a workshop.

III. To continue the exploration of value development after the workshop by developing the ability to know in which direction one's values are moving.

Group Size

Any number; participants basically work alone.

Time Required

Ten to fifteen minutes at the end of each session in a series of group sessions, or one-half hour at the end of each full day.

Materials

A pencil and a notebook for each participant.

Process

I. The facilitator explains the use and goals of the value-clarification journal, namely that the journal is used as a way to increase one's ability to intuit the inner movement of one's value formation—it is not merely a recording of facts and events. He may give a brief lecturette on how values develop and change (Rokeach, 1973, pp. 72-82).

II. (A) The facilitator instructs the members to write at the top of the left-hand page: "What happened in this session that stimulated me to reflect on the choosing, acting on, and prizing of a particular value?" He explains that the emphasis in this step is on recording what occurred in the session—the content learnings.

(B) The facilitator instructs the members to write at the top of the right-hand page: "What were my personal reactions (thinking and feeling) to the events of the session?" He explains that the emphasis in this step is on what each person learned about his own values during the session.

Adapted by permission of Hart Publishing Company, Inc., from its copyrighted volume VALUES CLARIFICATION: A Handbook of Practical Strategies for Teachers and Students by Sidney B. Simon, Leland W. Howe, and Howard Kirschenbaum.

(C) The facilitator instructs the participants to write at the bottom of the page: "My plan for growing in a particular value."

III. Participants are told that they will be given time at the end of each session to make notations under each heading.

Variations

I. A person who usually keeps a diary may set it up as a value-indicator record. On the left-hand sheet is written the usual diary entry. The right-hand sheet is used to record value-oriented reactions to the diary entry; this would consist of an evaluation of actions, wishes, desires, interests, beliefs, worries, use of time, use of money, use of energy, and so on, and of what these reactions helped the person to discover about his values.

II. A facilitator may choose to have participants complete the following sentences at the end of a session or workshop:

I have learned that I . . .

I realize that I . . .

I never knew that I . . .

I know now that I . . .

I have discovered that I . . .

I have relearned that I . . .

III. Using a chart such as the one below, a person who has kept a diary or journal for six months to a year may make a relative evaluation of the process of his growth in a particular value.

For similar techniques, see:

Pfeiffer, J. W., & Jones, J. E. (Eds.). *A handbook of structured experiences for human relations training,* 1974, *3* (74), pp. 109-111.

Progoff, I. *At a journal workshop.* New York: Dialogue House Library, 1975.

Raths, L. E., Harmin, M., & Simon, S. B. *Values and teaching.* Columbus, Ohio: Charles E. Merrill, 1966, pp. 139-140.

Simon, S. B. *Meeting yourself halfway.* Niles, Ill.: Argus Communications, 1974, pp. 20-21.

Simon, S. B., Howe, L. W., & Kirschenbaum, H. *Values clarification.* New York: Hart Publishing, 1972, pp. 163-165, 168-170, 388-391.

11. VALUE QUESTIONS

Goals

I. To acquaint the participants with the functional definition of a value.

II. To experience using the eight criteria in discovering one's values.

III. To distinguish values from value indicators.

Group Size

Any number of participants.

Time Required

Approximately one-half hour per value dealt with, or, as structured here, two to two-and-one-half hours.

Materials

(Facilitator should familiarize him- or herself with the "Introduction to a Value Dyadic Encounter" in technique No. 18 of this volume.)

I. Paper and a pencil for each participant.

II. Blank paper or copies of the Eight Basic Value Questions Worksheet for each participant.

III. A copy of the Value-Clarification Chart for each participant.

Process

I. Participants are instructed to list ten values they consider important. Specificity and detail are encouraged. (Five to ten minutes.) Then participants rank the list in terms of values considered most important. (Ten minutes.)

II. Each participant receives copies of the Eight Basic Value Questions Worksheet and is instructed to place a value on one of the sheets in the place provided. The value having the highest priority is considered first.

III. The facilitator asks the eight value questions and gives instructions to the group as follows:

1. Have you freely chosen this value? Where did the value come from? Do your parents hold this value? Answer these ques-

tions in the space provided. If you think you have freely chosen this value, write "Yes" in the space next to question 1.

2. From what alternatives did you choose this value? List the alternatives in the space provided.

3. List the consequences of the alternative you chose. Briefly, in short phrases, list the consequences you rejected in the other alternatives above. (If the participants are dealing with a complex value, they may need to use the back of the page to list the alternatives and consequences.)

4. How recently have you actually acted on this value? (Participants often list abstractions or qualities as values—this question helps them to state the value in terms of performance, e.g., "I value love" may be changed to "When and how did I last demonstrate my love for my spouse?")

5. Describe how this value has become a regular part of your life—a pattern or habit.

6. Give the date of the last time you publicly affirmed this value to someone.

7. How do you prize or celebrate this value in your life?

8. List the ways this value helps you to grow as a person: intellectually, morally, spiritually, etc.

(Groups will function at different paces, but this step may be expected to take from one-half hour to forty-five minutes.)

IV. Participants are instructed to take their second and third most important values from their rankings in step I and go through the same process on two other value sheets. (These rounds can usually be done in fifteen to twenty minutes.)

V. The participants are directed to consider their responses to the eight questions. They are reminded that unless all eight questions have been given some positive answer, the value is only a partial value or a value indicator.

VI. The facilitator divides the group into subgroups of eight persons each to share what each member has discovered about his values. The participants should share only what they feel comfortable sharing. (Forty-five minutes.)

VII. The activity may be concluded with a question-and-answer period, or the subgroups may list on newsprint the values they discussed and post them on the wall for all to see. The lists may have two columns: one for full values and one for partial values.

VIII. The facilitator hands out a copy of the Value-Clarification Chart to all participants, instructing them that this chart is for their own personal use and not necessarily to be shared with the group, unless the group members decide by consensus to do so.

Variations

I. With a group (or individual) that claims to have a given value, the Eight Basic Value Questions Worksheet may be used to determine whether it is a full value or a value indicator. If the value claimed is not a full value, the group members can explore alternatives, consequences, actions, and prizing criteria to help them develop the value they wish to hold.

II. The Value-Clarification Chart may be used with the facilitator asking the questions instead, as in step III.

Reference

Executive Council of the Episcopal Church. *A workshop on value clarification*. New York: Seabury Press, 1970, p. 43. (A booklet.)

EIGHT BASIC VALUE QUESTIONS WORKSHEET

Describe the value:

1. Have I freely chosen this value?

2. From among what alternatives? (List)

3. What are the consequences of choosing this value?

4. How recently have I acted on this value?

5. In what way has this value become a regular pattern in my life?

6. When did I most recently publicly affirm this value? (Give details)

7. How do I prize or celebrate this value in my life?

8. How does this value help me to grow as a person?

VALUE-CLARIFICATION CHART

(Note to group members: This chart is for your personal use and is not to be returned to the group leader.)

I. Write 1 or 2 key reminder words in answer to each statement below:

1. Something I'm saving money for:

2. The most important thing I did last year:

3. Something I did last year and do not do now:

4. What I would do for my next vacation if I could choose any possibility:

5. One thing I am proud of:

6. Something I own; a source of pleasure when someone mentions it:

7. Something difficult I'm glad I did:

8. Something I have learned to do in the past year:

II. Now, for each statement number, check the appropriate boxes below, E.g., if what you are saving money for is something you have chosen freely, check the first column box for that statement, and so on.

| | CHOOSING . . . | | | PRIZING . . . | | ACTING ON . . . | |
	Freely	From among alternatives	After thoughtfully considering each alternative and its consequences	Cherished	Affirmed publicly	Doing something with it	Repeatedly, as part of my life pattern
1.							
2.							
3.							
4.							
5.							
6.							
7.							
8.							

12. VALUE INTERVIEW

Goals

I. To better understand the values of another person.

II. To discover how another person has chosen, acted on, and prized a value.

III. To legitimize talking about one's values in a group.

IV. To open up new options to persons in a group.

Group Size

I. A larger group can use the group-on-group technique for the Value Interview by observing the facilitator with one of the group members.

II. For a getting-acquainted activity, the ideal group size is twelve to twenty.

III. With a larger group, the experience is best achieved by forming dyads and having the two persons take turns. Group members then form groups of eight to share the experience.

Time Required

Variable, depending on purpose. A group-on-group value interview may take from ten to twenty minutes. The processes described below may take from two to three hours.

Process

I. The facilitator conducts a value interview (Raths, 1966) for fifteen to twenty minutes, as a model for the group, utilizing the clarifying response technique (Raths, 1966) during the interview. (For a public interview, it is more functional for the group if the interviewee is strongly committed to a particular value.)

II. The facilitator may request that members of the group ask value-clarifying questions in an open forum. (Five to ten minutes.)

III. Participants are divided into groups of about eight members each; each group either agrees or disagrees with the interviewee's value. Each group selects a reporter and then explores the alternatives and consequences of the interviewee's value, as well as

Adapted by permission of Hart Publishing Company, Inc., from its copyrighted volume VALUES CLARIFICATION: A Handbook of Practical Strategies for Teachers and Students by Sidney B. Simon, Leland W. Howe, and Howard Kirschenbaum.

possible actions to be taken to implement the value. (Forty-five minutes to one and one-half hours.)

IV. Each group's reporter gives a three- to five-minute report. (Up to one-half hour.) If there are more groups than this time allows, the reporters may form a panel to share their groups' discussions for about one-half hour.

V. If desirable, the facilitator may instruct one half of each "pro" group to meet with one half of a "con" group for one-half hour to clarify and summarize their learnings.

VI. If members of a continuing group are keeping value-clarification journals, they may take ten to fifteen minutes to record their reactions and learnings.

Variations

I. The value interview may be used as an introductory activity (first to third session). With a group of ten to twenty persons, the interview is modeled for about twenty minutes, after which participants form dyads and interview each other for one-half hour. Members then volunteer to introduce the other member of the dyad to the group by telling about that person's values.

II. The value interview may be used as a brief, educational intervention. The interview is modeled for twenty minutes, then the group asks value questions for about five minutes. The group is broken into dyads to value interview each other for fifteen minutes, after which groups of eight are formed to share for one-half to one hour.

III. The value interview may be used when a dispute over a value arises in order to confront, and to try to resolve, the issue.

(A) In a group of twenty to thirty members, each side (pro and con) elects a representative. Each representative is interviewed separately for about ten minutes, with the focus on understanding the person's value position with regard to alternatives, consequences, and action. The two representatives conduct a value interview with each other, in turn, for about five minutes each and then go back to their respective groups to clarify the group's value position and to prepare to negotiate. This takes about one-half hour.

(B) The representatives come together to make an attempt at negotiation and reconciliation. The focus is on the difference in ranking priority and on exploring possible alternatives. Up to

four empty chairs may be placed alongside the two representatives in circular fashion so that members of either side may join
the representatives to give input. The negotiations continue until
a possible approach to conciliation is mutually agreed on.
(Twenty to forty minutes.)

(C) The representatives then go back to their groups to see
whether their members accept the proposal. If necessary, the representatives reconvene to make any changes or amendments
needed for group acceptance. Groups of eight, with four members
from each side of the issue, are formed to process this value activity. One-half hour is allotted for the processing.

For a discussion of value interviews and for similar techniques,
see L. E. Raths, M. Harmin, & S. B. Simon, *Values and teaching*.
Columbus, Ohio: Charles E. Merrill, 1966, pp. 142-152.

13. VALUE BRAINSTORMING

Goals

I. To explore and discover values through "value-laden" words.

II. To develop a "value awareness" of the attitudes expressed in the language and words used.

III. To encourage members of a group to be open about their values, thus making it easier for them to negotiate their value differences.

IV. To discover the direction or action a group may want to take.

Group Size

This activity is conducted with a large group, which is then broken into smaller groups of eight to twelve for sharing.

Time Required

Two to three hours.

Materials

Newsprint, felt-tipped pens, paper, pencils.

Physical Setting

The group is seated in a circle.

Process

I. The facilitator explains that the activity is a "brainstorming" experience. He defines brainstorming as a creative act that enables people to loosen up and associate ideas freely; in order to avoid blocking creativity, the items presented are not criticized or discussed during the brainstorming process. The facilitator then explains that when the imaginations of the group members are tapped, it is possible to gather the ideas and resources of the group as a unit.

Adapted by permission of Hart Publishing Company, Inc., from its copyrighted volume VALUES CLARIFICATION: A Handbook of Practical Strategies for Teachers and Students by Sidney B. Simon, Leland W. Howe, and Howard Kirschenbaum.

II. The group is invited to brainstorm the question "What are your primary concerns in life?" by calling out words, phrases, etc., that depict these concerns. While the members are brainstorming, the facilitator writes the words and phrases on newsprint for all to see. When the group pauses, he goes over the list with them to stimulate more ideas. This continues until the group has exhausted its input.

III. The facilitator describes how one's actual values may be masked by the words used to describe them. For example, if the list includes such words or phrases as "money," "college for the kids," "health," "family," "career advancement," "service to community," "good work relationships," etc., the facilitator may point out (beginning with the first word and proceeding down the list) how the word "money," for instance, may be indicative of such "hidden" values as security, power, self-worth, success, prestige, material comforts, and so on. The facilitator then writes these "hidden" values on newsprint so that the group members may evaluate them in terms of their relevance.

IV. The group members contribute their own evaluations of the value meanings of the words on their list. (For "money," for example, the most common responses seem to be family security, honesty, freedom, responsibility, material comfort, openness, happiness, courage, self-respect, forgiveness, sense of accomplishment, helpfulness, and capability.) The facilitator lists the group's responses and then compares the values in the two listings, collating the words that are similar. He notes if a particular value is repeated in different terms and points out that this may be an indicator of a core value of the group.

V. Each participant chooses four of the values listed on the newsprint and ranks them (first, second, etc.). The participants then form groups of four to six and share their values and rankings with one another. (About thirty minutes is needed for sharing.)

VI. Each small group is instructed to reach a consensus on a ranking of four values. In order that members of the group do not merely pool their values, no more than four ranked items are accepted. The members are also told to arrive at a common description or definition of each of the value words. (Thirty minutes.)

VII. The facilitator asks the members, "In light of the values that have emerged, what kinds of programs and activities could you be engaged in to affirm or act on those values?" He suggests that they evaluate whether actions they are currently engaged in agree with their values and with what they want to be doing. (Thirty minutes.)

VIII. The small groups report on their value discoveries and decisions. (Thirty minutes.)

Variations

I. This activity may be used as part of a planning or team-building process.

(A) If the activity is part of a planning process, a group size of six to ten is most effective and allows the facilitator to work closely with the group throughout the process.

(B) In a team-building workship, three hours is needed for the activity with another three hours allotted for the group to plan how it will achieve the goals it has decided on.

(C) Participants are instructed to describe or define what each of their four values (step V) means before they form smaller groups.

(D) Questions that may be asked in step VII above are:

1. "In light of your values, what activities should your organization be engaged in?"

2. "In light of your values, what kind of personal relationships should exist between employees and between management and employees?"

3. "In light of your values, what are the needs of your organization?"

4. "If there is a discrepancy between your values and the organization's product, procedures, relationships, activities, etc., what alternatives would conform to your values?"

II. This activity may be used as a first step in value discovery in conjunction with such other group dynamic techniques as problem solving, goal setting, conflict resolution, decision making, and planning. The questions in step VII may be modified to facilitate the particular goals for the group.

Reference

Hall, B. P., & Smith, M. *Value clarification as learning process: A handbook.* New York: Paulist Press, 1973, pp. 213-218.

14. VALUE PROBLEM SOLVING:
The Girl and the Sailor

Goals

I. To use the valuing process as a problem-solving structure.

II. To compare the value problem-solving approach with other problem-solving approaches.

III. To create an opportunity for group members to use one another as resource persons.

Group Size

Eight to twelve participants.

Time Required

Two to three hours. Varies according to problem and group.

Process

I. The facilitator gives a lecturette on the relation of the valuing process to the problem-solving process. He emphasizes the following points[1] in common, and notes the uniqueness of point 5, internalization of the value.

Valuing Process	*General Problem-Solving Process*
1. Need for choice	= Ask question or brainstorm
2. Identification of alternatives	= Identification of alternatives
3. Analysis of alternatives in light of consequences	= Establish criteria, rank alternatives, and synthesize
4. Choice among alternatives	= Conclusion
5. Internalization of the value	
6. Re-evaluation	= Re-evaluation

Adapted by permission of Hart Publishing Company, Inc., from its copyrighted volume VALUES CLARIFICATION: A Handbook of Practical Strategies for Teachers and Students by Sidney B. Simon, Leland W. Howe, and Howard Kirschenbaum.

[1]From D. S. Abbey, *Valuing*. Chicago: Instructional Dynamics, 1973.

II. The facilitator asks for five volunteers to participate in a group discussion. The volunteers form a circle in the center of the room; the remaining participants arrange themselves in an outer circle.

III. The facilitator tells the volunteer group members that he will present them with a problem that they must discuss and solve by consensus in twenty minutes. He tells them that they are to rank the five characters in a story in the order in which they appeal to the group. He informs them that he will interrupt the discussion at several points to allow the other participants to report their observations.

(These instructions are *not* repeated after the story is read.)

IV. The facilitator then reads the following story aloud.

The Girl and the Sailor[2]

A ship sank in a storm. Five survivors scrambled aboard two lifeboats: a sailor, a girl, and an old man in one boat; the girl's fiancé and his best friend in the second.

During the storm, the two boats separated. The first boat washed ashore on an island and was wrecked. The girl searched all day in vain for the other boat or any sign of her fiancé.

The next day, the weather cleared, and still she could not locate her fiancé. In the distance she saw another island. Hoping to find her fiancé, she begged the sailor to repair the boat and row her to the other island. The sailor agreed, on the condition that she sleep with him that night.

Distraught, she went to the old man for advice. "I cannot tell you what is right or wrong for you," he said. "Look into your heart and follow it." Confused but desperate, she agreed to the sailor's condition. The next morning the sailor fixed the boat and rowed her to the other island. Jumping out of the boat, she ran up the beach into the arms of her fiancé. Then she decided to tell him about the previous night. In a rage, he pushed her away and said, "Get away from me! I don't want to see you again!" Weeping, she started to walk slowly down the beach.

Seeing her, the best friend went to her, put his arm around her, and said, "I can tell that you two have had a fight. I'll try to patch it up, but, in the meantime, I'll take care of you."

---

[2]The story about the girl and the sailor is part of the folklore of value clarification. The particular version quoted here is from G. O. Charrier, "Cog's Ladder: A Process-Observation Activity," in J. W. Pfeiffer & J. E. Jones, *The 1974 annual handbook for group facilitators*, La Jolla, Calif.: University Associates, 1974, pp. 8-9.

V. Three or four times during the twenty-minute discussion the facilitator stops the action and asks the observers to comment on the ongoing valuing and problem-solving process and to "diagnose" the discussion in terms of value clarification. During this time, the five volunteers are asked to "bite the bullet" while listening to the observers.

VI. At the end of the twenty-minute discussion, the facilitator leads a discussion on the valuing process as a problem-solving process. He emphasizes the importance of internalizing values.

Variations

A number of group problem-solving activities can be used to illustrate the relationship between the valuing process and the general problem-solving process. The following listing refers the facilitator to twenty-five such activities (each with a detailed description of the process to be followed) in J. W. Pfeiffer and J. E. Jones, *A Handbook of Structured Experiences for Human Relations Training* (Volumes I-V, 1969-1975), and J. W. Pfeiffer and J. E. Jones, *The Annual Handbook for Group Facilitators* (1972, 1973, 1974, 1975, 1976). Each of these problem-solving activities has value implications. The facilitator may want to use the same process-interruption intervention described here to focus the issues involving values.

Sherwood, J. J. A Consensus-Seeking Task. In Pfeiffer & Jones, *A Handbook of Structured Experiences for Human Relations Training*, Vol. I, p. 49.

Goals: To compare the results of individual decision making with the results of group decision making; to teach effective consensus-seeking behaviors in task groups.

Time Required: Approximately one and one-half hours.

Pfeiffer, J. W. Choosing a Color: A Multiple-Role-Play. In Pfeiffer & Jones, *A Handbook of Structured Experiences for Human Relations Training*, Vol. I, p. 56.

Goals: To explore behavioral responses to an ambiguous task; to demonstrate the effects of shared leadership.

Time Required: Approximately forty-five minutes.

Residence Halls: A Consensus-Seeking Task. In Pfeiffer & Jones, *A Handbook of Structured Experiences for Human Relations Training*, Vol. I, p. 72.

Goals: To study the degree to which members of a group agree on certain values; to assess the decision-making norms of the group; to identify the "natural leadership" functioning in the group.

Time Required: Approximately one hour.

Group Tasks: A Collection of Activities. In Pfeiffer & Jones, *A Handbook of Structured Experiences for Human Relations Training*, Vol. II, p. 16.

Goals: To be used in studying group process.

Time Required: Varies with each activity.

Jones, J. E. NORC: A Consensus-Seeking Task. In Pfeiffer & Jones, *A Handbook of Structured Experiences for Human Relations Training*, Vol. II, p. 18.

Goals: To compare results of individual decision making and of group decision making; to generate data to discuss decision-making patterns in task groups.

Time Required: Approximately one hour.

Lutts and Mipps: Group Problem-Solving (based on Rimoldi). In Pfeiffer & Jones, *A Handbook of Structured Experiences for Human Relations Training*, Vol. II, p. 24.

Goals: To study the sharing of information in a task-oriented group; to focus on cooperation in group problem solving; to observe the emergence of leadership behavior in group problem solving.

Time Required: Approximately forty-five minutes.

Brainstorming: A Problem-Solving Activity. In Pfeiffer & Jones, *A Handbook of Structured Experiences for Human Relations Training*, Vol. III, p. 14.

Goals: To generate an extensive number of ideas or solutions to a problem by suspending criticism and evaluation; to develop skills in creative problem solving.

Time Required: Approximately one hour.

Kerner Report: A Consensus-Seeking Task. In Pfeiffer & Jones, *A Handbook of Structured Experiences for Human Relations Training*, Vol. III, p. 64.

Goals: To compare the results of individual decision making with the results of group decision making; to generate data to discuss decision-making patterns in task groups; to diagnose the level of development in a task group.

Time Required: Approximately one hour.

Supervisory Behavior/Aims of Education: Consensus-Seeking Tasks, (worksheets adapted from D. Nylen, J. R. Mitchell, & A. Stout). In Pfeiffer & Jones, *A Handbook of Structured Experiences for Human Relations Training*, Vol. III, p. 84.

Goals: To explore the relationships between subjective involvement with issues and problem solving; to teach effective consensus-seeking behaviors in task groups.

Time Required: Approximately one and one-half hours.

Twelve Angry Men Prediction Task. In Pfeiffer & Jones, *The 1972 Annual Handbook for Group Facilitators*, p. 13.

Goals: To compare the accuracy of predictions based upon group consensus seeking to those made by individuals; to generate data for a discussion of the merits of attempting consensus.

Time Required: Approximately two and one-half hours.

Energy International: A Problem-Solving Multiple Role-Play. In Pfeiffer & Jones, *The 1972 Annual Handbook for Group Facilitators*, p. 25.

Goals: To study how task-relevant information is shared within a work group; to observe problem-solving strategies within a group; to explore the effects of collaboration and competition in group problem solving.

Time Required: Approximately two hours.

Traditional American Values: Intergroup Confrontation. In Jones & Pfeiffer, *The 1973 Annual Handbook for Group Facilitators*, p. 23.

Goals: To clarify one's own value system; to explore values held in common within a group; to study differences existing between groups; to begin to remove stereotypes held by members of different groups.

Time Required: Approximately one and one-half hours.

Marion, D. J., & Edelman, A. Strategies of Changing: A Multiple-Role-Play (based on R. Chin & K. D. Benne). In Jones & Pfeiffer, *The 1973 Annual Handbook for Group Facilitators*, p. 32.

Goals: To acquaint people with three different interpersonal strategies for trying to effect change in human systems.

Time Required: Approximately one hour.

Zelmer, A. M. Shoe Store: Group Problem-Solving. In Pfeiffer & Jones, *A Handbook of Structured Experiences for Human Relations Training*, Vol. IV, p. 5.

Goals: To observe communication patterns in group problem solving; to explore interpersonal influence in problem solving.

Time Required: Thirty to sixty minutes.

Joe Doodlebug: Group Problem-Solving (adapted from M. Rokeach). In Pfeiffer & Jones, *A Handbook of Structured Experiences for Human Relations Training*, Vol. IV, p. 8.

Goals: To explore the effect of participants' response sets in a group problem-solving activity; to observe leadership behavior in a problem-solving situation.

Time Required: Approximately forty-five minutes.

Consensus-Seeking: A Collection of Tasks (worksheets by D. Keyworth, J. J. Sherwood, J. E. Jones, T. White, M. Carson, B. Rainbow, A. Dew, S. Pavletich, R. D. Jorgenson, & B. Holmberg). In Pfeiffer & Jones, *A Handbook of Structured Experiences for Human Relations Training*, Vol. IV, p. 51.

Goals: To teach effective consensus-seeking behaviors in task groups; to explore the concept of synergy in reference to outcomes of group decision making.

Time Required: Approximately one hour.

Dunn, L. Pine County: Information-Sharing. In Pfeiffer & Jones, *A Handbook of Structured Experiences for Human Relations Training*, Vol. IV, p. 75.

Goals: To explore the effects of collaboration and competition in group problem solving; to study how task-relevant information is shared within a work group; to observe problem-solving strategies within a group; to demonstrate the impact of various leadership styles on task accomplishment.

Time Required: Approximately one hour.

Joyce, J. L. Farm E-Z: A Multiple-Role-Play, Problem-Solving Experience. In Pfeiffer & Jones, *The 1974 Annual Handbook for Group Facilitators*, p. 44.

Goals: To study the sharing of information in task-oriented groups; to learn to distinguish a true problem from those which are only symptomatic; to observe problem-solving strategies within a group.

Time Required: Approximately two hours.

Inman, S. C., Jones, B. D., & Crown, A. S. Hung Jury: A Decision-Making Simulation. In Pfeiffer & Jones, *The 1974 Annual Handbook for Group Facilitators*, p. 64.

Goal: To study decision-making processes.

Time Required: Approximately two hours.

Nemiroff, P. M., & Pasmore, W. A. Lost at Sea: A Consensus-Seeking Task. In Jones & Pfeiffer, *The 1975 Annual Handbook for Group Facilitators*, p. 28.

Goals: To teach the effectiveness of consensus-seeking behavior in task groups through comparative experiences with both individual decision making and group decision making; to explore the concept of synergy in reference to the outcomes of group decision making.

Time Required: Approximately one hour.

Ford, D. L., Jr. Nominal Group Technique: An Applied Group Problem-Solving Activity (adapted from A. Delbecq & A. Van de Ven). In Jones & Pfeiffer, *The 1975 Annual Handbook for Group Facilitators*, p. 35.

Goals: To increase creativity and participation in group meetings involving problem-solving and/or fact-finding tasks; to develop or expand perception of critical issues within problem areas; to identify priorities of selected issues within problems, considering the viewpoints of differently oriented groups.

Time Required: Two hours.

Cash Register: Group Decision Making (based on W. V. Haney). In Pfeiffer & Jones, *A Handbook of Structured Experiences for Human Relations Training*, Vol. V, p. 10.

Goals: To demonstrate how decision making is improved by consensus seeking; to explore the impact that assumptions have on decision making.

Time Required: Approximately thirty minutes.

Sales Puzzle: Information Sharing (adapted from A. A. Zoll III). In Pfeiffer & Jones, *A Handbook of Structured Experiences for Human Relations Training*, Vol. V, p. 34.

Goals: To explore the effects of collaboration and competition in group problem solving; to study how information is shared by members of a work group; to observe problem-solving strategies within a group.

Time Required: Approximately one hour.

Joachim, J. R. Room 703: Information Sharing. In Pfeiffer & Jones, *A Handbook of Structured Experiences for Human Relations Training*, Vol. V, p. 39.

Goals: To explore the effects of collaboration and competition in group problem solving; to study how task-relevant information is shared within a work group; to observe group strategies for problem solving.

Time Required: Thirty to forty-five minutes.

Scott, K. D., & Pfeiffer, J. W. Letter Occurrence/Health Professions Prestige: Consensus-Seeking Tasks. In Pfeiffer & Jones, *A Handbook of Structured Experiences for Human Relations Training*, Vol. V, p. 44.

Goals: To compare decisions made by individuals with those made by groups; to teach effective consensus-seeking techniques; to demonstrate the phenomenon of synergy.

Time Required: Approximately one hour per task.

15. VALUE WORD SEARCH FOR COUPLES

Goals

I. To demonstrate the results of skillful communication in value clarification by identifying the value meaning underlying words.

II. To explore the implications of both marriage partners holding shared (or contrasting) meanings of key value words.

III. To increase experience both in clarifying one's feelings and in expressing them in a nonthreatening setting.

Group Size

Ten to thirty participants (five to fiteen actual or simulated couples).

Time Required

Approximately one hour.

Physical Setting

A comfortable room in which all participants can easily see one another.

Materials

I. A copy of the Value Word Search Worksheet for each participant.

II. Paper and pencils for all participants.

III. Tables, desks, or lapboards.

IV. Newsprint or a chalkboard (optional).

Process

I. After distributing the Value Word Search Worksheet and paper and pencils to all participants, the facilitator reads the following instructions to the group: "Working independently, find as many words on the Value Word Search Worksheet as you can in the allotted time. (Note examples already circled.) As you find a

Submitted by Barbara Jo Chesser and Ava A. Gray. Originally titled "An Experiential Approach to Value Clarification—Identification and Analysis of Values Through a Word-Search for Couple Simulation."

word, jot down the meaning that *first* occurs to you on the paper provided." (Approximately twenty minutes is allowed for the word search.)

II. When the first phase is complete, the facilitator asks the group to divide into simulated (or actual) couples to discuss the meanings that have been assigned to found words. Participants should discuss differences and similarities in their definitions, focusing on what implications their definitions have for their beliefs, values, goals, expectations, rules, and choices.

III. The group is re-formed and members are asked to draw conclusions about their findings using the following questions as springboards for discussion (the facilitator may list these questions on newsprint or on a chalkboard):

1. How do shared meanings contribute to a positive marital relationship? Do differences always suggest inevitable conflict? Give examples suggested by words on the Value Word Search Worksheet to support your answers.

2. Of what value is talking with a partner about your beliefs, goals, expectations, etc.?

3. How easy was it to listen nonjudgmentally to your partner's definitions? Does disagreement with your partner's definitions interfere with your acceptance of and respect for him or her?

4. How easy was it for you to share your definitions with your partner? How did his or her reactions influence the degree to which you were willing to take risks in self-disclosure?

5. In what ways do you feel skillful communication can facilitate increased understanding of your own and your partner's values? At what stage on the dating-marriage continuum does value clarification seem most critical?

Variations

I. Participants might find it informative and enjoyable to develop their own word search, incorporating words that have special significance to them.

II. For step III above, participants may work in subgroups of four, six, or eight when drawing conclusions about their findings. This variation would be especially useful if the entire group were so large that individual involvement and participation would be limited.

III. A videotape could be made of partners comparing and discussing their word definitions. The videotape could then be replayed so that participants can "read" the body language between themselves and their partners. A discussion could then take place concerning the congruence of body language with verbal communication.

IV. As a closure activity, participants may be instructed to write what they learned during the exercise. The content may be varied according to the needs and interests of the participants.

VALUE WORD SEARCH

```
W D A T I N G B R E L A T I O N S H I P S
B U D A G W R J Q N O C D D L T E E M E F
L M J H Y L O V E D N K S U A J X E U T N
D Q U P F M O I R U L H Z I L P B N K T V
C V S D G S M A R R I A G E O T C C Q I S
A O T I L T U B H I N P R D F P S H C N E
R O M A N C E L N N E P O E H A B A E G T
I Z E M O X V E W G S I W E D D I N G R T
N S N X I N C O M E S N T G D F K T L M I
G H T M A T U R I T Y E H O N E Y M O O N
E A Y C L M M G H I J S T U N Q S E M N G
A R B Q F R I E N D S S U V S R W N O E G
I I K N U S C O N F L I C T Y B X T Z Y O
J N N O P A P A R T N E R S F G A A B P A
Z G O C Q R R F G H M N M P H F I N C D L
D T U V H W O R K O S T V O L A J K D E S
I W W X Z I B X E W U R Q P T M Y L M N O
V B I R T H L V A L U E S K L I N L A W S
O F F K L M E D Y B S D E I J L O P Q R S
R G E J F D M A R E S P E C T Y Y N T U V
C R I S E S S W E E T H E A R T S Z S W X
E H I E C C O M M U N I C A T I O N A B C
```

16. VALUE-CLARIFICATION SHEET

Goals

I. To enable participants to clarify value issues.

II. To enable participants to share their values and resources.

III. To deal with environmental, social, and interpersonal values rather than with personal, individual values.

Group Size

Any number of participants.

Time Required

I. One and one-half hours for a simple technique.

II. Three hours for a more involved application.

Materials

Pencils and copies of Sample Worksheets I, II, III, and IV for each participant.

Process

I. Early in the value-clarification process, a group will develop its own list of value issues. From this, the facilitator (or the group) selects a topic and designs a value-clarification sheet that contains:

1. A strong provocative statement that will focus on the issue and stimulate the participants. (This may be a quotation from another source.)

2. Several value-clarifying questions that will facilitate the participants' exploration of their value stances.

3. A planning or problem-solving structure that will encourage decisions about possible actions to be taken.

II. The facilitator distributes the sample worksheets and participants take twenty to thirty minutes to reflect and to write their responses on the sheets.

III. Each participant takes about ten minutes alone to reflect on his experiences of the processes and on what he has learned about his position on the value issue.

Variations

I. In step III, homogeneous subgroups may be formed according to who agrees or disagrees with the value issue. The subgroups then elect representatives who will use the group-on-group technique to resolve the conflict. (Forty-five to sixty minutes.)

II. (A) Participants may work individually on their plans for action on the value issue (one-half hour) and then share these plans and use one another as resource persons (one-half hour).

(B) The group members may work together on possible actions to be taken; an activity such as Value Problem Solving may be utilized.

III. A short film may be used in conjunction with this technique.

———————

A similar activity appears in L. E. Raths, M. Harmin, and S. B. Simon, *Values and teaching*, Columbus, Ohio: Charles E. Merrill, 1966, pp. 83-111.

SAMPLE WORKSHEET I:
Rollo May's View of Our World Today

The striking thing about love and will in our day is that, whereas in the past they were always held up to us as the *answer* to life's predicaments, they have now themselves become the *problem*. It is always true that love and will become more difficult in a transitional age; and ours is an era of radical transition. The old myths and symbols by which we oriented ourselves are gone, anxiety is rampant; we cling to each other and try to persuade ourselves that what we feel is love; we do not will because we are afraid that if we choose one thing or one person we'll lose the other, and we are too insecure to take that change. The bottom then drops out of the conjunctive emotions and processes—of which love and will are the two foremost examples. The individual is forced to turn inward; he becomes obsessed with the new form of the problem of identity, namely, Even-if-I-know-who-I-am, I-have-no-significance. I am unable to influence others. The next step is apathy. And the step following is violence. For no human being can stand the perpetually numbing experience of his own powerlessness.[1]

1. Does Rollo May's statement agree with your experience of the last ten years? If not, change the statement so that it does agree with your experience. If the statement agrees with your personal experience, write about it.

2. How do you see the process: a person loses his ability to love and will and so moves from an identity search to apathy, to feeling powerless, to violence? Are there other possibilities in this dynamic? List them.

3. List the variables that need to be changed in order that a person will not end up in apathy and violence.

4. List the conditions that help a person to grow to his potential.

---

[1]Reprinted from *Love and Will* by Rollo May. By permission of W. W. Norton & Company, Inc. Copyright © 1969 by W. W. Norton & Company, Inc.

SAMPLE WORKSHEET II:
Carl Rogers on the Goal of Education

The world is changing at an exponential rate. If our society is to meet the challenge of the dizzying changes in science, technology, communications, and social relationships, we cannot rest on the *answers* provided in the past, but must put our trust in the *processes* by which new problems are met. In the world which is already upon us, the aim of education must be to develop individuals who are open to change. . . . The goal of education must be to develop a society in which people can live more comfortably with *change* than with *rigidity*. In the coming world the capacity to face the new appropriately is more important than the ability to know and repeat the old. But such a goal implies, in turn, that educators themselves must be open and flexible, effectively involved in the process of change.[2]

1. If you disagree with the above statement, change it so that you can agree with it. If you agree with the statement, give your reasons for agreeing.

2. List all the possible goals of education that you can think of.

3. Choose three of your goals from item 2 and list the consequences of adopting each as your goal for education.

4. List all the factors that would have to be changed in order to achieve the highest goal you have for education.

5. Make a plan for how you might effect change in the present educational system to arrive at your goal for education.

---

[2]From *Freedom to Learn: A View of What Education Might Become* by Carl Rogers. Columbus, Ohio: Charles E. Merrill Publishing Company, 1969, pp. 303-304.

SAMPLE WORKSHEET III: Dealing with Conflict

Conflict is a daily reality for everyone. Whether at home or at work, an individual's needs and values constantly and invariably come into opposition with those of other people. . . . The ability to resolve conflict successfully is probably one of the most important social skills that an individual can possess. . . . Conflict-resolution strategies may be classified into three categories: avoidance, defusion, and confrontation. . . . Successful negotiation, however, requires a set of skills which must be learned and practiced. These skills include (1) the ability to determine the nature of the conflict, (2) effectiveness in initiating confrontations, (3) the ability to hear the other's point of view, and (4) the utilization of problem-solving processes to bring about a consensus decision.[3]

1. If you disagree with the above statement, change it so that you can agree with it. If you agree with the statement, write a brief description of an experience from your life that illustrates it.

2. List other possible strategies besides avoidance, defusion, and confrontation. Describe or define what these three terms mean to you, along with definitions of your own terms.

3. Describe how you handle conflict at work or at home.

4. Compare your way of dealing with conflict with the skills suggested above. Which, if any, of these skills do you need to develop or improve?

5. Choose a conflict that is important to you or to your small group and either (a) role play the situation and use the small-group members as resource people to improve your skills, or (b) use a problem-solving approach and use the small-group members as resource persons.

---

[3]From "Conflict-Resolution Strategies" by Joan Stepsis. In Pfeiffer, J. W., & Jones, J. E. *The 1974 annual handbook for group facilitators*, La Jolla, Calif.: University Associates, 1974, pp. 139-141. Used with permission.

SAMPLE WORKSHEET IV:
Emotional Support in Marriage

In summary, marriage used to be an institution for the *physical* survival and well-being of two people and their offspring. Today, in our industrialized and affluent society, couples are not struggling to exist physically as in the past. Instead, we have primarily the struggle for *psychological* and *emotional* survival. However, the changes in the structure, form, and processes of marriage have been too few and too unsystematic to cope with the new psychological and emotional needs and problems of today. All too often today, couples invest their emotional needs in their children instead of in each other. Then, when the children grow up and leave, the couple discover that there is no emotional relationship between them.[4]

1. Do you agree with the above statment? If not, change the statement so that you can agree with it. Explain why you agree or disagree with the statement.

2. List three things that are important for your psychological and emotional survival.

3. What three things are preventing you from attaining psychological and emotional support in your marriage?

4. List the values you and your spouse share. List the values you do not share. What keeps you from sharing them?

5. List the activities you do together as a couple (exclude activities with the children). List activities you would like to do together.

---

[4]Reprinted from *The Mirages of Marriage* by William J. Lederer and Dr. Don D. Jackson. By permission of W. W. Norton & Company, Inc. Copyright © 1968 by W. W. Norton & Company, Inc.

17. THE VALUE OF FRIENDSHIP

Goals

   I. To explore the value of friendship.

  II. To develop awareness of one's relationship to others.

 III. To develop friendships within the group.

Group Size

    Any number of participants.

Time Required

    Two to three hours.

Materials

    Pencils and copies of Worksheets I and II for all participants.

Process

   I. (A) If a group has shown an interest in exploring the value of friendship, love, relationships with others, intimacy, etc., the facilitator begins with an activity on "phone numbers."

     (B) He hands out Worksheet I and says that each participant is to list all of the people with whom he has talked on the telephone during the past two weeks. If a participant does not know the name of someone with whom he has spoken, he is instructed to write "salesman," "customer," etc. (Fifteen minutes.)

  II. Members are instructed to:

    1. Write all the telephone numbers that they know from memory in the number column next to the name of the person with whom they spoke.

    2. Put an "A" in the number column if the person's telephone number is in their personal address book.

    3. Put a "P" to indicate those numbers that they would have to look up in the phone book.

    (These notations take from five to eight minutes.)

 III. The facilitator gives the following instructions, listing the symbols on newsprint as he does so. (Passing out copies of the instructions would reduce the effectiveness of the experience.)

1. Write the letter "E" in column 1 next to the names of the first three people you would call in an emergency.
2. Write the letter "G" in column 2 next to the names of people you would call if you had good news.
3. Write the letter "C" in column 3 by the names of people you would call for comfort in sadness.
4. Write the letter "L" in column 4 after the names of the people with whom you talk the longest in a given week.
5. Write the letter "I" in column 5 next to the names of the people who have influenced you the most.
6. Write the letter "F" in column 6 next to the names of the people with whom you have the most fun.
7. Write the letter "Q" in column 7 next to the name of one person whom you wish would stop calling you.
8. Write the letter "B" in column 8 by the names of two people whom you consider to be your best friends.

IV. The facilitator allows five minutes for participants to evaluate their lists individually. He then asks:
1. What interesting patterns do you see?
2. What name has the most marks?
3. Where did you meet these people?
4. Were there any surprises for you in doing this activity?

V. Participants are directed to mingle and then to form groups of about eight persons each, in which they will feel comfortable sharing what they have learned about themselves. (Forty-five to sixty minutes.)

VI. Participants then evaluate how they might improve their relationships with their friends. (Twenty minutes.)

VII. The small groups reconvene and members use one another as resource persons and exchange suggestions about their plans.

Variations

I. (A) The facilitator introduces the "Circle of Friends" activity. He directs each participant to put his name in the center of a sheet of paper and to put the names of his friends in a circle around his name, with the people he feels closest to near his own name and the names of the people he is not as close to farther away. (Ten minutes.)

(B) The facilitator tells the participants to draw a line between the names of their friends who know one another. He asks:

1. Do you have a small circle of friends who all know one another or are your friends more dispersed?
2. What qualifications must a person have in order to become one of your friends?
3. What were the qualities of these friends that attracted you to them?
4. What are some of the things your friends like to do together?
5. When was the last time you did something with your friends?
6. Did all enjoy it?
7. What are the differences among your friends?

(C) Participants form groups of about eight people with whom they feel comfortable sharing their "circle of friends." They discuss what they might be able to do for their friends.

II. (A) The facilitator describes the "Whom Do You Trust?" activity and gives the participants five to ten minutes to list their closest friends and associates.

(B) He passes out Worksheet II and allows twenty to thirty minutes for participants to complete the worksheet.

(C) Participants form groups of about eight persons each to share their worksheets. (Forty-five to sixty minutes.)

(D) Participants take about fifteen minutes to plan how they will improve their relationships in some way or how they will do something they have been meaning to do for a friend.

Reference

Hartwell, M., & Hawkins, L. (with Simon, S. B.). *Value clarification: Friends and other people*. Arlington Heights, Ill.: Paxcom, 1973, pp. 34-42, 58-64, 106-114.

WORKSHEET I: Phone Numbers

NAMES	NUMBERS	1.	2.	3.	4.	5.	6.	7.	8.

Adapted from Hartwell, M., & Hawkins, L. (with Simon, S. B.). *Value clarification: Friends and other people*. Arlington Heights, Ill.: Paxcom, 1973, p. 41.

WORKSHEET II: Whom Do You Trust?

Of all the people you listed as your closest friends, whom would you trust to:

_____ 1. Keep a very important secret for a year?

_____ 2. Share a very serious problem?

_____ 3. Substitute for you when you cannot do something to which you have committed yourself?

_____ 4. Accurately carry an important message to another friend?

_____ 5. Give you good advice when you have a disagreement with your boss or supervisor?

_____ 6. Help you with a problem at work?

_____ 7. Get an important item for you by a deadline (a book, medicine, your glasses, etc.)?

_____ 8. Call you to let you know about an activity in which you would be interested?

_____ 9. Act as arbitrator if you had a disagreement with your spouse or a very close friend?

_____ 10. Meet with a visiting friend if you were unable to be there?

1. Which of the above choices were the hardest?—the easiest?

2. Which items would your best friend ask of you?

3. List the criteria you have for deciding you can trust someone.

4. How do other people know they can trust you?

5. What did you learn from this activity?

6. What can you do to improve your relationships with your friends?

Adapted from Hartwell, M., & Hawkins, L. (with Simon, S. B.). *Value clarification: Friends and other people*. Arlington Heights, Ill.: Paxcom, 1973, p. 113-114.

18. A VALUE DYADIC ENCOUNTER

Goals

I. To facilitate sharing one's values with another person on an intimate level.

II. To learn the use of value-clarifying responses in order to help another person clarify his own values.

Group Size

Any number of dyads.

Time Required

I. Approximately one hour to study the value-clarifying responses prior to the activity.

II. A minimum of ninety minutes for the encounter. The time allowed should be open-ended.

Materials

I. A copy of the Introduction to a Value Dyadic Encounter for each participant.

II. A copy of criteria for the Use of Value-Clarifying Responses for each participant.

III. A copy of Hints for Beginning the Clarifying Response for each participant.

IV. A copy of A Value Dyadic Encounter Booklet for each participant.

V. A copy of the Self-Evaluation Form for Clarifying Responses for each participant.

Process

I. The Introduction to A Value Dyadic Encounter (including three lists of value-clarifying responses), the Criteria for the Use of Value-Clarifying Responses, and the Hints for Beginning the Clarifying Response are given out at least twenty-four hours prior to the activity. The participants are instructed to study these materials and try to memorize all the responses.

II. To begin the activity, the facilitator distributes a copy of A Value Dyadic Encounter Booklet to each participant. Both members of each dyad open their booklets at the same time and follow the instructions in the booklet. The facilitator informs the participants of the length of time available for completion of the activity.

III. At the completion of the activity, the facilitator hands out a Self-Evaluation Form to each participant. Before filling out the form, each participant receives feedback from his partner and considers such feedback in his total self-evaluation. Completed forms may be shared with the facilitator.

IV. The process, significant learnings, and other reactions may be discussed by the total group.

For similar programmed encounters, see:

Pfeiffer, J. W., & Jones, J. E. (Eds.). *A handbook of structured experiences for human relations training* (Vol. I). La Jolla, Calif.: University Associates, 1974, pp. 90-100.

Pfeiffer, J. W., & Jones, J. E. (Eds.). *A handbook of structured experiences for human relations training* (Vol. II). La Jolla, Calif.: University Associates, 1974, pp. 97-100.

Pfeiffer, J. W., & Jones, J. E. (Eds.). *A handbook of structured experiences for human relations training* (Vol. III). La Jolla, Calif.: University Associates, 1974, pp. 89-93.

Pfeiffer, J. W., & Jones, J. E. (Eds.). *A handbook of structured experiences for human relations training* (Vol. IV). La Jolla, Calif.: University Associates, 1973, pp. 45-48, 66-74.

Pfeiffer, J. W., & Jones, J. E. (Eds.). *A handbook of structured experiences for human relations training* (Vol. V). La Jolla, Calif.: University Associates, 1975, pp. 116-130.

Thayer, L. (Ed.). *Affective education*. La Jolla, Calif.: University Associates, 1976, pp. 46-54.

INTRODUCTION TO A VALUE DYADIC ENCOUNTER

We live in a time of great moral confusion. Countless issues in recent history indicate the value dilemmas which we face . . . women's rights; rights of blacks, chicanos and other minority groups; abortion; organ transplants, students' rights; aging; marriage; energy and environ-

---

Submitted by Bernard Nisenholz, Richard Tirman, and Frederick McCarty. Used with permission.

mental problems; and countless others.

With the growth of technology and the new advancements in science, life becomes more and more complex. Simple answers on what one chooses to value become more difficult to find. Traditional approaches to value development such as inspiring, preaching, setting rules and regulations, cultural or religious dogmatizing, or appealing to one's conscience do not seem to work well for many. More and more people seem to be experiencing alienation and value confusion in their lives.

A humanistic approach to value development views values as very personal things, the deepest level of value being one's self value. In order for one's self value to emerge, an atmosphere that provides an individual with choices, freedom, and responsibility is necessary. A humanistic approach to values, therefore, would see value development as a process very similar in nature to the process of self-actualization.

The valuing process includes the following criteria:

CHOOSING — ACTING — PRIZING

1. *Choosing* implies having free choice, being aware of alternatives, and considering the consequences of those alternatives.
2. *Acting* implies acting on one's choices, taking responsibility for one's actions, and acting in a regular pattern.
3. *Prizing* implies feeling proud of one's choices, sharing them with others, accepting one's self regardless of the choices, and enhancing one's self (growing) because of the choice.

One way to help people understand their values is through the use of certain value-clarifying responses. These responses stimulate people to clarify their choosing, prizing, and acting in order to become more self-directed.

Please study carefully the following three lists of clarifying responses that you will be using in A Value Dyadic Encounter.

CLARIFYING RESPONSES
CHOOSING

1. What would be the payoff for you?
2. What would you get out of your behavior?

3. Where would that idea lead; what would be its consequences?
4. What are some good things about that notion?
5. What other possibilities are there?
6. Have you considered some alternatives?
7. How do you correct what you did?
8. Who tells you that you should?
9. Is that what you really want?
10. What do you see in front of you?
11. What is the worst thing that could happen to you?
12. If you really felt completely free to choose without any other forces to consider, what would you do?
13. What prevents you from choosing that?
14. Did you really choose that?

CLARIFYING RESPONSES
ACTING

1. Would you really *do* that?
2. Can you take responsibility for that?
3. Can I help you do something about your idea?
4. Would you do the same thing over again?
5. Do you value that?
6. How do you prevent yourself from doing that?
7. What does your intuition tell you to do?
8. Have you checked your feeling out with whomever the feeling concerns?
9. When will you do that?
10. Would you like to make a contract to do that?
11. What are your expectations of catastrophe if you do that?
12. How do you start?
13. What is one thing you can do right now?
14. What do you do about that?

CLARIFYING RESPONSES
PRIZING

1. Can you say that your experience is O.K.?
2. Is that something you feel good about?
3. Are you glad about that?
4. Would you share that with anyone?
5. Was that nourishing?
6. Was that self-enhancing?
7. What are you feeling?
8. "When you say that, I feel . . ." (Share *your* feelings about the person's statement.)
9. How does that make you feel?
10. Would you tell (him/her/them) how you feel about that?
11. How has that helped you grow as a person?
12. Is that helping you to develop in the direction you want?
13. You seem to like that, is that correct?
14. Are you dissatisfied with what happened?

CRITERIA FOR THE USE OF VALUE-CLARIFYING RESPONSES

An effective clarifying response encourages a person to think through something he has done or said. It is part of a process to help participants clarify their values.

There is no guarantee that any one type of response will produce the desired effect, but the following elements help to identify responses that are likely to do this. The elements outline the criteria used to assess the effectiveness of a clarifying response.

1. Did the questioner try to promote his own values? Was praise, criticism, or moralizing used to guide the other person's thinking?
2. The response puts the responsibility on the person to look at his ideas or his behavior and thus to think and decide for himself what it is he wants. The other person (or dyadic partner) must be free to choose his own values.

3. Was the emotional climate conducive to thoughtfulness? A respectful accepting attitude is usually best. Fear, guilt, embarrassment, or other forms of emotional distress clog clear thinking. The main focus must be on the other person's perceptions and not on the questioner's.
4. Was the clarifying response gentle enough? It should be stimulating but not insistent. A participant should comfortably be able to decide not to think about the issue.
5. Was the interaction brief enough? With too many clarifying comments, the other person has so much to think about that he may tend to dismiss it all. A clarifying response does not try to do big things with brief exchanges. It sets a mood of thoughtfulness. The effect is cumulative. The role of the questioner is that of assisting the other person to see more widely and broadly his entire experiential field so that the whole spectrum of values and meanings becomes open and visible to him.
6. Were the responses more to obtain data than to stimulate thinking? Clarifying responses are not intended for interview purposes. One listens carefully, not primarily for information, but for possible gaps in the clarifying process.
7. Was the clarifying response used on an appropriate topic? It is best for attitudes, feelings, beliefs, purposes—it is not for facts. It is not appropriate when one has in mind a correct answer for the other person to give. The clarifying response is not for drawing the other person toward a predetermined answer.
8. Was the response used sensitively and creatively? Clarifying responses are not mechanical things that follow a formula. They must be used creatively, to fit oneself and the situation. When a response helps a client to clarify his thinking or behavior, it is considered effective. The three lists of clarifying responses are given as examples. They are not to be used mechanically or slavishly.

HINTS FOR BEGINNING THE CLARIFYING RESPONSE

The clarifying response is one of the central techniques of value clarification and is often found to be the most difficult to master. One of the reasons for this is probably that it requires a person to make substantial changes in his style of interacting with others.

The following hints are helpful in becoming more adept at using the clarifying response:

1. *Pause*. Some questioners have difficulty with clarifying responses because they assume that their response must be immediate. This helps make them tense and uncertain when attempting to use it. Actually, people appreciate it if the questioner seems to be considering their remarks more carefully.

Accept the other person's response (perhaps by nodding or saying something like "I understand," pause and take time to think, then respond.

2. *Be Yourself*. The clarifying response strategy may seem to imply that you have to *erase* your normal pattern of responses and substitute another, foreign system. It would be foolish to attempt to erase your normal human responses in this way, and it probably would not work, anyway.

Example:

Dyadic Partner: "I'm going to spend my college money on a new car."

Questioner: "Really?" (Pause. During the pause, the questioner thinks what he would ordinarily say. Perhaps this would be "You'll regret that later," or "That seems like a foolish choice, Bill." Instead of saying this, the questioner considers what he is trying to accomplish. In this case, it would be to get the other person to consider the consequences of such a choice. He then uses a clarifying response to achieve this.)
"Have you considered the consequences of doing that, Bill?"

A VALUE DYADIC ENCOUNTER BOOKLET

The following ground rules should govern this experience:

The discussion items are open-ended statements that can be completed at the level of self-disclosure one chooses to use.

Everything discussed should be kept strictly confidential.

Do not look ahead in the booklet.

Each partner responds to each discussion item before continuing.

The items are to be completed in the order in which they appear.

You may decline to respond to any item by simply telling your partner that you pass, but do not skip items.

You may take whatever time you wish with each item.

Look up. If your partner has finished reading, turn the page and
begin.

- -

My name is . . .

- -

My job is . . .

- -

Right now I feel . . .

- -

(Please read silently and follow the instructions.) Review the list of
value-clarification responses that deal with *choosing*. Review the items
carefully, and then use the responses with your partner when completing the next set of statements. When responding, do not look at the list,
but you may resort to it if you are stuck for a response.

- -

1. Something I would really like to do is . . .

- -

2. Something I'm really afraid of doing is . . .

- -

3. The most confusing value issue in my life is . . .

- - - - - - -   -

4. What I would really like to tell my (boss/spouse/
   parents) is . . .

- -

5. One thing I've always wanted to do but have
   not is . . .

- -

6. A difficult decision I face is . . .

- -

7. The options in the above (#6) decision are . . .

- -

8. If only I could . . .

- -

9. The way I feel about smoking marijuana
   is . . .

- -

10. The reason why I feel I (should/should not)
    engage in (extramarital/premarital)
    sex is . . .

- -

(Please read silently and follow the instructions.)

Discuss the exchanges during the previous clarifying responses and share your feelings.

- -

(Read silently and follow the instructions.)

Review the list of value-clarification responses that deal with acting. Read them carefully and then use the responses with your partner when completing the next set of statements. When responding, do not look back at the list.

- -

1. If I had a magic wand that I could wave to make
   any three changes at (home/work/recreation),
   I would ...

- -

2. As far as smoking is concerned, I ...

- -

3. As far as (insert a social issue that implies
   a value stance) is concerned, I ...

- -

4. Five years from now I see myself doing ...

- -

5. A behavior I would like to change is ...

- -

6. One value (or behavior) I would never
   change is ...

- -

7. Something I'm firmly convinced of now, but
   was not (5/10/15) years ago is ...

- -

8. If I knew I was going to die in a month and
   could do anything I wanted during
   that month, I would ...

- -

9. The hardest thing about my life is ...

- -

10. When I meet someone I am sexually
    attracted to, I ...

- -

(Please read silently and follow the instructions.)

Discuss the exchanges during the previous clarifying responses and share your feelings.

- -

(Read silently and follow the instructions.)

Review the list of value-clarification responses that deal with *prizing*. Read them carefully and then use the responses with your partner when completing the next set of statements. When responding, do not look back at the list.

- -

1. Something that I really bungled recently is . . .

- -

2. I am really good at (something you feel
   proud of) . . .

- -

3. The most serious lie I have ever told is . . .

- -

4. Someone in my family I am proud of is . . .

- -

5. Something I just learned is . . .

- -

6. A difficult decision I made recently was . . .

- -

7. I have (difficulty/no difficulty) in telling
   you about . . .

- -

8. Something that helped me to grow recently
   was . . .

- -

9. I am proud of my position on (insert a
   value you stand by) . . .

_ _

10. One thing I do not tell others about is . . .

_ _

(Read silently and follow instructions.)

On a separate sheet of paper, write two incomplete statements. The
first incomplete statement is for you to respond to and the second
incomplete statement is for your partner to respond to.

Share your experience of this total exercise. Check the self-evaluation
form and discuss your responses to each other.

OR

Check the evaluation form on the basis of your feelings about your
partner's responses. Then share your self-evaluation forms with other
group members, discussing how you feel about others' responses on the
form.

SELF-EVALUATION FORM FOR CLARIFYING RESPONSES

	Extremely	Very	Acceptably	Slightly	Not At All
1. Was it a clarifying situation?					
2. Did exchanges end gracefully?					
3. Was I as natural as I could be?					
4. Did my own values get in the way?					
5. How comfortable did I feel using this strategy?					
6. Did I notice changes in my interaction styles?					
7. How accepting was I of my partner's responses?					
8. How well did I remember some specific clarifying responses?					
9. How good was my timing; did I pause briefly, before responding to my partner?					
10. How comfortable would I feel using these strategies again?					

19. POLARIZING POSITIONS

Goals

I. To use creatively the opportunity to learn from someone who strongly disagrees with a popular value.

II. To promote a third position in a group consisting of those who hold a value and those who do not hold the value.

III. To increase the possible range of alternatives from which to choose a value.

Group Size

Twenty participants work together best, but the technique can be successful with larger groups.

Time Required

One and one-half to two hours.

Materials

Newsprint and felt-tipped pens may be used for listing.

Process

I. The facilitator identifies a person (or persons) in the group who genuinely opposes an otherwise popular value and invites him to present his views to the group. The facilitator legitimizes the advocate's position by requesting openness on the part of the group and by presenting the advocate's role as an opportunity for group members to be exposed to a different viewpoint.

II. The advocate takes about fifteen minutes to state his views, including how he arrived at his value, the alternatives and consequences he considered, what actions he has taken, and how much he prizes his value.

III. Participants form subgroups of about eight persons each and discuss their reactions to the advocate's discourse. (Forty-five to sixty minutes.)

IV. One-half hour is allowed for the advocate to respond to questions and to discuss issues with the total group.

V. Members take ten minutes for individual reflection on the experience. If they are keeping value-clarification journals, they may work on them during this period.

VI. The facilitator takes approximately fifteen minutes to process the total experience with the group and to lead a discussion with members about what they have learned from the experience.

Variations

I. If there is no one in the group who holds an opposing opinion on a popular value, the facilitator may assume the role of devil's advocate or he may ask a participant to assume that role and proceed as outlined above. The person selected should be good at role playing and must understand the points of view of the other participants.

II. Polarized value positions may be the subject of a value interview conducted by the facilitator as step II.

III. In place of step II, if there is a minority of three to six persons who oppose the group consensus these people form an inner group in a group-on-group discussion of choosing, acting on, and prizing their value. Then, in step IV, three of the minority-view advocates and one member from each subgroup may discuss the issue within a group-on-group arrangement. (To be effective, the discussion should consist of no more than eight persons.)

Reference

Raths, L. E., Harmin, M., & Simon, S. B. *Values and teaching*. Columbus, Ohio: Charles E. Merrill, 1966, pp. 127-128.

20. INTERNATIONAL POLITICS

Goals

    I. To simulate the value differences present in international negotiation.

    II. To develop group cohesiveness and self-awareness.

    III. To increase skill in setting goals, negotiating, and resolving conflicts.

Group Size

Twenty to thirty participants; members divide into four or five subgroups.

Time Required

One and one-half to two hours.

Materials

    I. Chalkboard or newsprint and markers.

    II. Paper and pencils for all participants.

    III. A copy of Worksheet I for each participant.

Process

    I. The facilitator divides the group into four or five subgroups and then gives the following instructions:

        1. Your subgroup is the executive planning board of a new nation, and your first task is to set values for relative power and wealth.

        2. Consider the *amounts* of power and wealth that your nation will need. (You will be given vehicles for attainment of some or all of these values.)

        3. Some things to consider when setting values are:

Power	*Wealth*
Defensive power	National wealth
Aggressive power	Individual standards of living
Peace	Wealth as power

Submitted by Jay Irwin Levinson.

Domestic tranquility
Territorial stability
Power to help other nations

4. Each subgroup is to choose a spokesperson who will also serve as negotiator.

II. Subgroups are told that they have fifteen minutes to establish and write down the power and wealth goals for their individual nations. They are also told to set one world-wide power and wealth goal. These are the goals they will try to attain. At the end of the time period, each subgroup publicly announces its goals.

III. The facilitator distributes copies of Worksheet I, which gives the rules concerning war, negotiating, and dependency.

IV. The facilitator randomly determines the power and wealth distribution by placing a number for each group (nation) in a container and having a representative from each group draw one number. The facilitator then distributes Value Units according to the numbers drawn (as shown below) and explains the meaning of the Value Units. These meanings (Section B below) are outlined on newsprint or on the chalkboard by the facilitator.

A. Value Unit Distribution

1. If Four Groups:
Group I   —Four power units.
Group II —Three wealth units.
Group III—Two wealth units.
Group IV—Three wealth units.

2. If Five Groups:
Distribute as above but give the extra group one power unit and two wealth units.

B. Meaning of Value Units

1. *Power Units*

Zero units   = No territorial, civil, international stability.

One unit   = Peace against outside aggression and inner (civil) stability.

Two units   = Power for any purpose. However, the possessor of this Value Unit cannot aggress against a nation that possesses peace (one unit).

More than
two units   =   Useless except for trade negotiation.

2. *Wealth Units*

Zero units   =   Poverty; population is literally
starving.

One unit   =   Minimum life sustenance.

Two units   =   Middle-class fluctuating economy.

Three units   =   Wealth.

Four units   =   Money enough to buy a power unit
(extra power unit may be added to game.)

V. The facilitator instructs groups to begin their negotiating session(s) with one spokesperson representing each nation. The negotiations take place in a group-on-group arrangement, with the negotiators in the center. Each session is a maximum of fifteen minutes; the facilitator enforces the time rule.

VI. Immediately before the negotiating session begins, Group I is instructed that it may not trade away more than two of its power units.

VII. During the activity, questions should be asked of the facilitator in private by the individual groups, as opposed to the facilitator asking for questions from the group as a whole. This will allow for more intragroup interaction and value judgment. The facilitator should give the groups as much freedom as possible to devise their own solutions to the problems that arise.

VIII. Any number of outcomes is possible. For example, in one instance, the experience ended in Group I declaring war on all the other nations. Negotiations broke down, so war was declared. Since none of the other groups had any power units, the control group conquered all and "won" the game.

In another session, Group I ignored the rules (using the power they were given), and traded away all but one power unit because it conformed to their world-wide goal (peace for all nations).

In the first instance, although everyone said that they wanted peace, the activity ended in war. In the second instance, the world-wide goal took precedence over the rules of the game.

IX. The experience ends when, in the opinion of the facilitator, no further negotiation is possible and/or probable. Groups are advised to stay within the time frames stated above for negotiating sessions. Two negotiating sessions are maximum. War may also end the game.

X. The facilitator leads a discussion on the experience, focusing on the valuing process employed by the subgroups. He may also want to emphasize that the negotiating process has intrapersonal implications in addition to intergroup processes.

Variations

 I. Facilitator can allow negotiators to confer with their nations during Step V.

 II. Observers can be instructed to take notes on value processes they observe during the negotiating session(s).

WORKSHEET I: Rules for Negotiating and for Declaring War or Dependency

Power	Units	
Zero units	=	No territorial, civil, international stability.
One unit	=	Peace against outside aggression and inner (civil) stability.
Two units	=	Power for any purpose. However, the possessor of this Value Unit cannot aggress against a nation that possesses peace (one unit).
More than two units	=	Useless except for trade and negotiation.

Wealth

Zero units	=	Poverty; population is literally starving.
One unit	=	Minimum life sustenance.
Two units	=	Middle-class fluctuating economy.
Three units	=	Wealth.
Four units	=	Money enough to buy a power unit (extra power unit may be added to game).

I. Negotiating

1. There may be one or two negotiating sessions, which may be started anytime after each group's goals have been set.
2. Each session should be no more than fifteen minutes.
3. Groups may want a feedback session between sessions to determine how far along they are in their goal attainment.
4. Groups may decide to limit the negotiations to one session depending on how interesting the outcome of the first session is. *Observers should be used.*

II. Declaring War

War may be declared anytime after approximately two-thirds of the first negotiating session at the discretion of the leader. If you allow war to be declared too quickly, the game may break down prematurely.

III. Declaring Dependency

A nation can only become *dependent on* another nation. Merging is not permitted. To become dependent on another nation, you must:

1. give up all assets to the mother nation.
2. be at the ruling mercy of the other nation.
3. have benefit of *their* assets as agreed.

All power and wealth unit values are now in terms of double population; that is, *two* power units now constitute peace, etc.

21. LIFE RAFT

Goals

I. To simulate as closely and dramatically as possible the experiencing of a value, rather than merely the intellectualizing of it.

II. To identify the feelings involved in a particular value.

III. To confront the participants' intellectualizations with an experiential understanding of the value.

Group Size

Ten participants per facilitator.

Time Required

One and one-half to two hours.

Materials

A loud, manual alarm clock.

Process

I. (A) The facilitator instructs the group to sit in the middle of the floor, in a formation resembling a life raft. He sets the scene by asking the participants to imagine that they have been on an Atlantic cruise, that a serious storm has developed, that their ship has been struck by lightning, and that they have all had to get into a life raft. He explains that the major problem that now exists is that the raft has room and food enough for only nine persons and there are ten in the group. One person must be sacrificed in order to save the rest. The facilitator informs the group that the decision is to be made by group consensus: Each member is to "plead his case" to the others, arguing why he should live, and then the group is to decide who must go overboard. He tells the participants that they have one-half hour to make their decision. At the end of that time, the life raft will sink if there are still ten people in it. He puts a manual alarm clock near the participants so that they can hear it tick and sets the alarm to go off in one-half hour.

(B) At intervals during the decision-making process, the facilitator notifies the group of the time remaining.

II. The facilitator leads the group in processing the dynamics and the feelings that have emerged during the activity. Since the experience is powerful, sufficient time (one-half hour or more) must be allowed to complete this task successfully.

III. The facilitator then directs the group to brainstorm the values that are implicit in the situation they have just experienced. He asks the following questions:

1. What kind of value assumptions did members of the group make?
2. What values were the members acting on?
3. What did you learn about your values from an experiential standpoint?
4. In light of this experience, how do you value your own life and the lives of others?
5. What is your worth?

Variations

I. More than one person can be "sacrificed" in the life raft.

II. The values of love and charity may be explored through "Coins: Symbolic Feedback" (Pfeiffer & Jones, *A Handbook of Structured Experiences for Human Relations Training*, Vol. I, p. 104).

III. The values of cooperation and collaboration may be explored through "Consensus Seeking: A Collection of Tasks" (Pfeiffer & Jones, *A Handbook of Structured Experiences for Human Relations Training*, Vol. IV, p. 51).

IV. The value of competition and the participant's theology/ philosophy of man may be explored through "Win as Much as You Can: An Intergroup Competition" (Pfeiffer & Jones, *A Handbook of Structured Experiences for Human Relations Training*, Vol. II, p. 62).

V. The values of clear communication, of avoiding judgments, and of dealing with facts may be explored through "Rumor Clinic: A Communications Experiment" (Pfeiffer & Jones, *A Handbook of Structured Experiences for Human Relations Training*, Vol. II, p. 12).

VI. The values of power and influence may be explored through "Line-Up and Power Inversion: An Experiment" (Pfeiffer & Jones, *A Handbook of Structured Experiences for Human Relations Training*, Vol. III, p. 46).

References

Raths, L. E., Harmin, M., & Simon, S. B. *Values and teaching*. Columbus, Ohio: Charles E. Merrill, 1966, pp. 123-124.

Sax, S., & Hollander, S. *Reality games: Games people "ought to" play*. New York: Popular Library, 1972.

Simon, S. B. *Meeting yourself halfway*. Niles, Ill.: Argus Communications, 1974, pp. 66-70.

22. VALUE INDICATORS: TIME/MONEY

Goals

I. To use the value indicator of time to discover an individual's values.

II. To explore the development and formation of values so that one's behavior will become congruent with one's values.

III. To have individuals reflect on the direction in which they want to grow in terms of their values.

Group Size

Any number of participants.

Time Required

One and one-half to two hours.

Materials

A copy of Worksheet I, II, or III and a pencil for each participant.

Process

I. The facilitator states that the group will explore the value indicator of time. He passes out copies of Worksheet I and directs participants to rank the categories of work, family, recreation, and culture according to which they value most. They place the numbers 1 through 4 in the column marked "Rank." The ranking takes from three to five minutes, depending on the size of the group.

II. The facilitator instructs participants to reflect on what they do with their time in a typical week. He points out the blank spaces in each category and says that these may be filled in to suit an individual's life style. The total times listed should account for 112 hours in a typical week (168 hours − 56 hours for sleeping = 112). (Thirty minutes.)

III. Participants then compare their initial rankings with the actual time spent on each item.

IV. The facilitator gives a lecturette on the ways in which time is an indicator of values. He notes that time spent thinking about work is counted as work time and that leisure does not truly begin until a person has recuperated from work.

V. The group shares discoveries resulting from this experience. (Forty-five minutes.)

VI. The facilitator may allow fifteen minutes for participants to plan individually how they will adjust their time to meet their values. This plan may involve a major overhaul of time allotment or may require merely a more efficient use of time in order to allow time for the person to do the things he says he wants to do.

VII. Participants then regroup and use one another as resource persons to discuss their plans.

Variations

I. For a more loosely structured activity, the participants may be directed to fill in a twenty-four-hour clock face according to how they use their time in a typical day, week, or month. A "Pie of Life" may be used as an example.

II. Money is another value indicator. Many people keep track only of their total income and expenditures and have no real idea about how or where they spend their money. Worksheet II is used to help participants estimate how they spend money and then plan how they would like to change these patterns. Participants are encouraged to check their worksheets against their bankbooks when they return home.

III. Worksheet III uses the value indicator of time to chart work time versus leisure time to determine if a participant truly values leisure. The worksheet also specifies with whom the time is spent—self, family, etc. Participants are again reminded that work time includes travel, planning, and worrying about work, as well as time spent recuperating from work. A person must have the energy to invest in leisure before he can enjoy it. In other words, a person who comes home from work and watches television is likely to be recuperating from work rather than indulging in recreation.

References

Hall, B. P. *Value clarification as learning process: A guidebook.* New York: Paulist Press, 1973, pp. 128-129, 254-255.

Hall, B. P. *Value clarification as learning process: A sourcebook.* New York: Paulist Press, 1973, pp. 99-100, 145.

Hall, B. P., & Smith, M. *Value clarification as learning process: A handbook.* New York: Paulist Press, 1973.

Raths, L. E., Harmin, M., & Simon, S. B. *Values and teaching.* Columbus, Ohio: Charles E. Merrill, 1966, pp. 139-140.

Smith, M. Some implications of value clarification for organization development. In J. E. Jones & J. W. Pfeiffer (Eds.), *The 1973 annual handbook for group facilitators.* La Jolla, Calif.: University Associates, 1973.

WORKSHEET I: Value Indicator—Time

Activities	Rank	Time Spent
Work Planning or worrying Performing Recuperating Other _____ Other _____		
Family Time with spouse Time with children Time with total family Time with friends by yourself Time with friends with your spouse Time with friends with your family Other _____ Other _____		

Adapted from Smith, 1973, pp. 4-5.

Activities	Rank	Time Spent
Recreation Reading Sports Movies Television Conversation Parties Time by yourself Hobbies _____ Other _____ Other _____		
Culture Performing arts Church Political activities Community involvement Other _____ Other _____		

WORKSHEET II: Value Indicator—Money

Item	Money Spent	Rank in Order of Cost	Rank as You Would Prefer
Sustenance Food Mortgage or rent Lodging/renovations or repairs Clothes Transportation Medical Other _____ Other _____			
Development Study (books, courses, etc.) Mental and physical health Recreation (including equipment) Travel Hobby (including equipment) Other _____ Other _____			

Adapted from B. P. Hall, *Value clarification as learning process: A sourcebook.* New York: Paulist Press, 1973, p. 99.

Item	Money Spent	Rank in Order of Cost	Rank as You Would Prefer
Future Insurance Pension Savings Profit sharing Stocks Other _____ Other _____			
Miscellaneous Liquor Vacations Gambling Parties Other _____ Other _____			

WORKSHEET III: Value Indicator—Time

Hours Spent With	Self	Spouse	Family	Work Colleagues	Friends	Other (clubs, church, etc.)	Time Totals
Hours spent during a typical week at work obligations, duties, expectations, responsibility							
Hours spent during a typical week in maintenance and rest or recuperation							
Hours spent during a typical week in leisure							
Time Totals:							

Total time to account for: 112 hours (7-day week—56 hours for sleep).

Adapted from B. P. Hall & M. Smith, *Value clarification as learning process: A handbook.* New York, Paulist Press, 1973, p. 193.

23. VALUE ROLE PLAY

Goals

I. To explore dramatically and in depth the implicit values in a relationship between two or more people.

II. To discover the values that are indicated by feelings.

III. To explore the value impositions that frequently exist in conflicts between people.

Group Size

Fifteen to twenty people.

Time Required

Varies, depending on group and role-play situation.

Materials

Newsprint and felt-tipped pens (optional).

Process

I. The facilitator determines an issue within the group that may be clarified through role playing. For example, a group of married couples may discuss the high cost of college educations for their children and then talk about their teen-agers' lack of appreciation for the use of money and for the opportunity to go to college. The facilitator invites one couple to role play their feelings and selects two volunteers to play the couple's teen-age children.

II. The facilitator describes a role for each participant or tells the participant to play himself. In the case described above, the roles might be:

1. *Father*—You love your kids and you want to send them to college, but you feel unappreciated because they do not seem to be interested in going to college.

2. *Mother*—You love your family and agree with your husband that the kids should go to college. You cannot understand why they do not seem to appreciate his desire to give them a better education.

3. *Son*—You are seventeen years old, a senior in high school, and you do not want to go to college. You are not sure what you

131

want to do but you like to work with your hands and think you would like to become an electrician, machinist, or carpenter.

4. *Daughter*—You are sixteen years old, a junior in high school, and do not want to go to college. You would just like to be married, but do not know what to do in the meantime.

III. The facilitator or participants set the scene, e.g., around the table after dinner. Participants assume their positions and begin to role play.

IV. During the role play, the facilitator may whisper directions or suggestions to any of the players. Other participants are encouraged to offer their suggestions in the same manner. If the role players introduce another character into the discussion, e.g., Uncle Jim, who favors college education, another participant may assume that role and enter the scene.

V. The facilitator stops the role play at an appropriate time, when underlying values, feelings, and attitudes have surfaced. The role players are not expected to solve the problem.

VI. The facilitator leads the group in processing what happened during the role play. The discussion includes any understandings or misunderstandings that were formulated.

VII. The values implicit in the role-play situation are identified by the participants. The facilitator may list on newsprint the values that emerged for each role.

Variation

In light of the value-clarification discussion, the role players may be asked to assume their roles once again to see if they can resolve their value differences.

For a discussion of value role playing, see L. E. Raths, M. Harmin, & S. B. Simon, *Values and teaching*, Columbus, Ohio: Charles E. Merrill, 1966, pp. 121-123.

24. VALUE-INDICATOR SEARCH

Goals

I. To learn how to use value indicators to discover one's values.

II. To learn the difference between a partial value and a full value.

III. To discover emerging values and to plan further value development.

Group Size

Unlimited number of six-person groups.

Time Required

Two to three hours.

Materials

A pencil and a copy of Worksheet I or paper for each participant.

Process

I. The facilitator delivers about a ten-minute lecturette on value indicators.

II. The facilitator hands out a copy of Worksheet I to each participant and says that each person is to choose one of his values that he would like to develop further and write that value at the top of the worksheet. Participants then answer the questions on the worksheet. (Twenty to thirty minutes.)

III. Participants form subgroups of six persons each and use other members as resource persons to share their worksheets and help one another formulate ways to develop their particular values. (Approximately one hour.)

IV. Each participant devises a plan for a way to grow in the value he has listed.

V. If participants know one another fairly well, the facilitator may choose to rearrange the subgroups so that each is composed of persons who are working on similar values. In this way, they may function more effectively as resource persons for one another, exploring the alternatives and the consequences of the values in question and the actions that may be taken. (Forty-five to sixty minutes.)

133

Variation

(A). The facilitator describes how behavior manifests values. He models the discovery of values by selecting a volunteer who names one of his values. The group then brainstorms a list of observable behaviors that would indicate this value. For example, if the value is "family togetherness," the value-indicating behaviors might be:

1. I speak up against business, church, and club activities that are for men only or women only and I promote activities that are family oriented. (An expression of an attitude.)

2. Someday we will have a pool in the backyard. (An aspiration.)

3. We plan to have a Sunday picnic twice a month during the summer. (A goal.)

4. My spouse and I read family-oriented magazines. (Carrying out an interest.)

5. Every Saturday afternoon we have a family discussion and then play a game together. (An activity.)

6. I often dream of buying a house (farm) where my children can have houses (farms) nearby. (A daydream.)

7. We budget forty-five dollars every two weeks for family recreation, such as dining out, bowling, etc. (Use of money.)

8. Our family has an hour-long supper together every evening, during which we discuss whatever is on our minds. (Use of time.)

9. I am often concerned that some harm may come to one of my children. (A worry.)

10. We like to take family vacations with the children. (A preference.)

(B). The volunteer participant checks to see whether these behavioral indicators apply to him. The more similarities he finds, the more likely it is that his value fulfills the eight criteria for a full value. The other participants check their behavior against the list to see whether a value emerges.

(C). After the facilitator has modeled three or four such behavior explorations, the participants form subgroups of eight to ten persons and continue the exploration.

WORKSHEET I: Value-Indicator Search Questions

The value: _____

Give yourself one point for each question for which you can answer yes or can list the activity mentioned.

_____ 1. What have you done to promote this value?

_____ 2. What is your opinion about this value if it is positive?

_____ 3. If you have chosen this value over something else, identify the rejected alternative.

_____ 4. Do you have a long-range plan to implement this value?

_____ 5. Do you ever say, "One of these days, I'm going to (the value)?"

_____ 6. Do you ever daydream about this value?

_____ 7. Do you have a date set to act on this value in the near future?

_____ 8. Do you subscribe to magazines about or read books about this value?

_____ 9. If this value fits in with one of your goals of life, identify that goal.

_____ 10. What would you spend a weekend doing that manifests this value?

_____ 11. Have you ever tried to convince someone else to act on this value?

_____ 12. Have you ever had a dream at night about this value?

_____ 13. How much money do you spend on this value?

_____ 14. How much time do you spend on this value in a month?

_____ 15. How much time do you spend per month thinking or worrying about this value?

List five specific observable behaviors that would demonstrate that you perform this value:

16. _____

17. _____

18. _____

19. _____

20. _____

Relative scale: 20-16: probably a value; check the criteria for a value

15-11: on the way to being a value

14- 6: the beginning of a value

5- 0: not a strong value

References

Hall, B. P. *Value clarification as learning process: A guidebook*. New York: Paulist Press, 1973a.

Raths, L. E., Harmin, M., & Simon, S. B. *Values and teaching*. Columbus, Ohio: Charles E. Merrill, 1966, pp. 30-33, 65-72.

25. CONVERSION OF LIMITATIONS

Goals

I. To discover the positive value inherent in an apparently negative limitation.

II. To explore attitudes about positive and negative values.

III. To increase one's feeling of self-worth.

Group Size

Eight to twelve people per facilitator.

Time Required

Approximately one and one-half hours.

Materials

Paper and a pencil for each participant.

Process

I. Fifteen minutes is allotted for participants to make a list of those attitudes and/or activities that they consider to be limitations.

II. The facilitator tells participants to take another fifteen minutes to consider their lists and to search for positive values in the items listed. For example, if a person has listed that he becomes angry when he is interrupted at work, the positive value may be his dedication to or involvement and interest in his work. Another example might be a person who feels a limitation because he is unable to repair his own car and hates being dependent on automobile mechanics. The underlying positive value here may be the person's desire to be in control of situations and a desire for increased knowledge and skills.

III. Each facilitator takes a subgroup of eight to ten participants and helps them to share their lists and to use one another as resource persons in converting their limitations to positive values.

IV. Participants take ten minutes to write about how they feel as a result of the activity. (Value journals may be used for this step.)

Variations

I. (A) The activity may be called "talents and limitations" and may be used to enable a person to evaluate himself in terms of his role as a friend, employer, teacher, leader, etc. It also may be used to explore relationships with others. The facilitator tells participants to take a sheet of paper and make two vertical lines to create three equal columns. Then they list their talents in the first column and their limitations in the second column.

(B) Participants use the third column to convert their limitations to positive values. For example, a person who says a limitation is that he does not socialize well at parties may identify the underlying positive value as his desire to have only close relationships with a few people.

(C) Participants form subgroups of eight to ten people to share their lists and to help one another convert limitations to positive values. In an intact group, members may add items to one another's lists.

II. In an organization development setting, participants may unite to list the positive aspects and the limitations of the organization as a system. They then attempt to convert the limitations to positive values. If this cannot be done, they may want to reconsider the listing.

References

Hall, B. P. *Value clarification as learning process: A guidebook*. New York: Paulist Press, 1973, p. 169.

Hall, B. P. *Value clarification as learning process: A sourcebook*. New York: Paulist Press, 1973, pp. 49-50.

Smith, M. Some implications of value clarification for organization development. In J. E. Jones & J. W. Pfeiffer (Eds.), *The 1973 annual handbook for group facilitators*. La Jolla, Calif.: University Associates, 1973, pp. 203-211.

26. FANTASY EXPLORATION

Goals

I. To use fantasy as a means of discovering values.

II. To understand how fantasies can be an indicator of values.

III. To share one's values in a more personal context with the other group members.

Group Size

Any number of participants.

Time Required

Approximately one and one-half hours, preferably in the evening.

Materials

A copy of Worksheet I and a pencil for each participant.

Physical Setting

A carpeted room large enough for all participants to lie down on the floor. The facilitator must be able to control the lights.

Process

I. The facilitator describes how fantasy is an indicator of values and then invites the participants to accompany him on a fantasy trip.

II. (A) The facilitator tells the participants to lie down in a comfortable position on the carpet and to close their eyes. He tells them to become aware of their breathing and to breathe deeply in and out. He may use any relaxing exercise he chooses or may merely continue to talk in a soothing voice . . .

> "As you breathe in and out, let yourself become relaxed, become aware of your body, explore your body. If you feel tension in any muscle, tighten the muscle, hold it, tense it, hold it. Now relax it, and as you relax let the relaxation flow throughout your body. Continue to breathe deeply, in and out. Make sure you are in a comfortable position. Breathe in and out . . ."

(B) The facilitator continues:

> "Imagine that you are in the woods. What kind of woods is it? What kind of trees do you see? What does the sky look like? Do you see it in

color? Let the woods come into focus. Is there anything other than trees in the woods? Bushes? Flowers? Animals?

"Look into the near distance; there is a house in the woods. What does the house look like? Explore the outside of the house. What do you imagine the inside of the house looks like? Go in. What is inside the house? What are you doing while you are inside the house?

"Come out of the house now. Look ahead; there is a path. What kind of path is it? Where do you think the path leads? You are walking down the path; what does it feel like? Just a few feet away, you notice there is a paper cup lying by the path. What do you do with the cup? All of a sudden, you see a snake crossing the path right in front of you. What do you do? What kind of snake is it? Now the snake is gone. Continue walking down the path. You begin to see a small body of water ahead of you. Go up to the water. What does it look like? What feeling do you get from the water? What color is it? Is anything reflected in the water? Then turn and continue walking on the path. You notice that there is a key lying just ahead of you. What kind of key is it? What do you do with the key?

"You continue down the path and come to a curve. Around the bend, you see a wall blocking the path. You know that the path continues on the other side of the wall. What kind of wall is it? You want to continue on the path; what do you do about the wall? How did you do it? Once you are on the other side of the wall, what do you see? What does the other side of the wall look like? What else do you see?"

(C) The facilitator may modify the fantasy trip as he sees fit, leaving all descriptions general and nonspecific so that the participants may fill in with their own imaginations. He continues to speak slowly as he tells participants that the fantasy trip has ended and instructs them to take a minute or two to open their eyes and sit up comfortably. The above process takes from twenty-five to thirty minutes.

III. The facilitator passes out Worksheet I and briefly explains the code, avoiding controversy over specific items. He gives the participants ten to fifteen minutes to record what they imagined on the fantasy trip.

IV. Participants form groups of approximately six persons each and share their fantasy trips with each other. (Thirty minutes.)

V. The facilitator directs the participants to adopt a value-clarification attitude and to explore with one another the values implicit in their fantasy trips. He says that they need not accept the Freudian or Jungian interpretation of the symbols, but that the general, total context does have a value meaning for everyone. This exploration and sharing may take as long as forty-five to sixty minutes.

Variations

I. This fantasy is adapted from psychosynthesis. The participants lie down and close their eyes. The facilitator helps them to relax by employing any relaxation exercise he feels comfortable in using. If the group seems to be tense, the Yoga method of tightening and tensing each major muscle and then letting that part of the body flop down in relaxation is very helpful. The facilitator tells the participants to begin with the right leg, hold it about eight inches off the floor, tense the muscle, hold it for a minute, then let it flop down on the rug. They raise the other leg, each arm separately, lift the buttocks, shoulders, and head in a similar way. Then the facilitator begins the fantasy of climbing the mountain to meet the wise old man.

"Imagine that you are in a valley and that off in the distance you can see a snow-capped mountain . . . See the mountain as clearly as you can. What does the sky look like? . . . Notice that there is a path leading to the mountain. Walk toward the mountain. See the path going up the mountain . . . Begin to ascend the mountain. How do you go up the mountain? . . . You are about a third of the way up; what do you see from this vantage point? . . . Begin to anticipate that you are going to find something very interesting. You are about two-thirds of the way up the mountain. From this viewpoint you can see that there is a flat area at the top of the mountain and you begin to move toward it . . .

"As you reach the top of the mountain where the flat spot is you see that there is a bench and someone sitting on it. Instantly you realize that it is a very wise old man. You feel comfortable and safe in approaching him. Look closely at the wise old man, notice his clothes, head, face, nose, mouth, eyes, etc. . . . Sit down next to him. You know that you can ask any question you want and he will be able to answer you. Form a definite question in your mind . . . Ask your question; be silent, be still; listen, listen to the wise old man." (The facilitator leaves the group in silence for five to fifteen minutes until he perceives that people are beginning to stir.) "You are at the top of the mountain. You have been listening to a very wise old man. It is time to leave. Say good-bye, take one final look, and turn and walk away. The descent down the mountain is smooth and fast. As you reach the half-way mark you begin to see that there are other people who have been up to the top of the mountain. You continue down and all meet together in the valley around a camp fire . . .

Take a minute to slowly open your eyes and sit up at a comfortable pace."

II. Invite the participants to recall a dream they have had and to relive the dream in their fantasy. Some will say that they do not dream; invite them to relive a daydream in fantasy. Follow a similar structure as above to explore the values.

III. Once a group has had some experience with guided fantasy they are much more open to fantasizing on their own. Invite a group that has had some experience with guided fantasy to spend about one-half to one hour to fantasize where they will be and what the world will be like in twenty years. Have them share their fantasies with one another for about one-half hour. Then ask them to adopt a value-clarification attitude and explore the value implications of their fantasies.

References

Otto, H. *Fantasy encounter games*. New York: Harper & Row, 1972.

Pfeiffer, J. W., & Jones, J. E. (Eds.). *A handbook of structured experiences for human relations training* (Vol. I). La Jolla, Calif.: University Associates, 1974, pp. 75-79.

Pfeiffer, J. W., & Jones, J. E. (Eds.). *A handbook of structured experiences for human relations training* (Vol. IV). La Jolla, Calif.: University Associates, 1973, pp. 92-94.

Pfeiffer, J. W., & Jones, J. E. (Eds.). *The 1976 annual handbook for group facilitators*, La Jolla, Calif.: University Associates, 1976, pp. 191-201.

WORKSHEET I: Ten-Symbol Fantasy Path

Note to participant: Keep the context of your total fantasy in mind as you attempt to discover the values implicit in your fantasy. You do not have to accept this interpretation of the symbols; only you can make a valid interpretation of your fantasy. Take about ten minutes to record your fantasy using this code sheet.

WOODS = your life:

OUTSIDE OF HOUSE = how others see you:

INSIDE OF HOUSE = how you see yourself:

PATH = your future:

PAPER CUP = the way you treat strangers or other people:

SNAKE = the way you react to danger:

WATER = your emotional life:

KEY = what you do with knowledge:

WALL = death:

THE OTHER SIDE OF THE WALL = the afterlife:

27. VALUES AND CONFLICT

Goal

To experience the resolution of conflict through the clarification of value priorities (a) with couples and larger families, and (b) in an organizational system.

Group Size

Twenty or more participants (including a process observer for each of four subgroups of four persons).

Time Required

Two to three hours.

Materials

    I. Newsprint and felt-tipped pens.

    II. A dollar in small change (including as many pennies as possible) for each participant.

    III. Masking tape.

Process

    I. (A) The facilitator introduces participants to the Ajax Widgets Company and tells them that they will experience the effects of value ranking within an organization. He writes on newsprint the words "function," "ideas," "feelings," and "imagination." He explains that these represent the four major areas of personality and role. *Function* is getting the job done, accomplishing the task. *Ideas* encompass the intellectual approach—getting and understanding the facts. *Feelings* means sensitivity to other persons and friendship. *Imagination* is the searching approach, as well as creativity.

    (B) The facilitator tells the participants to rank the four words individually, according to which is most important to each of them as a person and as a worker.

    II. Participants mingle and choose partners. Each set of partners then selects another pair with whom to share and clarify their definitions of the four words. This discussion takes about twenty minutes.

III. The facilitator designates the four corners of the room as the four divisions of the Ajax Widgets Company: production, research, management, and sales. He instructs participants to go to the division (corner) in which they think they would like to work. If the groups turn out to be uneven in size, the facilitator may ask that some participants make a second choice in order to divide the company into fairly even sections.

IV. Each division defines its orientation by ranking the four words in their order of importance to the operations of that division. Each division lists its ranking on newsprint and posts it on the wall.

V. Each participant compares his initial personal ranking of the four words with the lists on the wall and joins a work division whose ranking most closely agrees with his own. The facilitator may need to even out the groups again by asking persons in larger divisions to make a second choice.

VI. The work groups take twenty minutes to discuss why they chose that particular division as well as what they think that division's task is within the company.

VII. The facilitator directs the participants to form subgroups composed of one person from each division plus a volunteer to serve as process observer. He informs the process observers quietly and individually that they are to watch for value conflicts arising from differences in rankings.

VIII. The groups are informed that their task is to develop a budget for the company for the next month. The company has $40,000 to spend. The money may not merely be divided equally among the four divisions, but is to be allocated on the basis of what each division contributes to the company. At this point, each of the four active members of each group contributes his dollar in change. One penny equals one hundred dollars, so the four dollars on hand represents the $40,000 to be budgeted. Participants are told that they have forty-five minutes to complete the task. If it is not accomplished by that time, the main office (the facilitator) will take back the available funds.

IX. When the time is up, the facilitator explains that conflict results from different priorities being given to the same values. He directs the participants to process how their different rankings of function, ideas, feelings, and imagination caused conflict in the decision on how to allocate the $40,000. The processing takes about one-half hour.

X. The facilitator leads the group in a discussion of the total experience. He emphasizes an awareness of how the experience can be applied to actual work dynamics.

Variation

The conflict-resolution technique may be used with married couples, larger families, or small work groups. The facilitator has the participants list their most important values and then rank them. They then compare the differences. This process will include a clarification of their true values, what they mean by the value terms, and what values are implicit in their words and actions.

References

Hall, B. P. *Value clarification as learning process: A guidebook*. New York: Paulist Press, 1973.

Jones, J. E., & Pfeiffer, J. W. (Eds.). *The 1975 annual handbook for group facilitators*. La Jolla, Calif.: University Associates, 1975, pp. 51-53, 56-62.

Pfeiffer, J. W., & Jones, J. E. (Eds.). *The 1972 annual handbook for group facilitators*. La Jolla, Calif.: University Associates, 1972, pp. 44-50.

Pfeiffer, J. W., & Jones, J. E. (Eds.). *The 1974 annual handbook for group facilitators*. La Jolla, Calif.: University Associates, 1974, pp. 22-23.

Pfeiffer, J. W., & Jones, J. E. (Eds.). *A handbook of structured experiences for human relations training* (Vol. I). La Jolla, Calif.: University Associates, 1974, p. 70.

28. NEGOTIATION OF EXPECTATIONS

Goals

I. To apply value clarification to organization development.

II. To negotiate expectations, set goals, and accomplish team building through value clarification.

Group Size

Twelve persons per facilitator.

Time Required

One to three hours, depending on the issues dealt with.

Materials

I. Newsprint and felt-tipped pens for each participant.

II. Paper and a pencil for each participant

III. Masking tape.

Process

I. The facilitator introduces the participants to the activity. He tells them that it is intended to reveal hidden agendas and, by working through them, to increase open communication. He tells participants that they are to make individual lists of what they want the organization to achieve. He emphasizes that they should list specific plans of action that may be achieved realistically and not deal in generalities, abstractions, gossip, or Utopian ideals.

II. Participants force rank their lists according to which performance should be accomplished first. Each participant writes out his ranking in large letters on a sheet of newsprint.

III. The rankings are hung on the wall and the group compares them and evaluates the direction prescribed for the organization and what role individuals play in working toward that direction. The members may discover that they need certain skills. They will also discover their strengths and interests.

IV. The facilitator leads the group members in clarifying their vision for the organization and in describing what they mean by such terms as collaboration, open communication, accountability, success, etc.

V. Each participant then lists his talents, skills, and experience. He force ranks the list on newsprint, according to which are his favorites, and then hangs the ranking next to his original "what to do" ranking.

VI. The group examines the listings to discover previously unknown assets. If a participant is not interested in pursuing a talent, he may be willing to consult with others about it. In this way, the group becomes aware of its resources.

Variation

(A) The facilitator informs participants that they will be establishing goals and priorities according to their values and that they must first clarify what they mean by the words they use to describe values. Participants list their goals for the organization, including both personal and corporate goals. Then, in large letters on newsprint, they set priorities for their goals by ranking them from first to last. They hang their rankings on the wall for all to see.

(B) The group checks the rankings to see which, if any, particular value emerges as most important. The facilitator determines whether all participants mean the same thing when they use this value word by asking them to write on newsprint what they mean when they use the word. He selects one participant to define the word as he uses it. He then chooses another participant to define the word in his own terms, and so on, until enough different definitions have been obtained to demonstrate the point.

(C) The facilitator helps the group members identify and clarify their underlying values by having them list and rank the values that underlie their goals and priorities. These lists are also hung on the wall.

(D) The group strives to reach consensus on the priority of its goals.

References

Hoy, T. A *values clarification design as an organization development intervention*. Washington, D.C.: Project Test Pattern, 1973.

Jones, J. E., & Pfeiffer, J. W. (Eds.). *The 1973 annual handbook for group facilitators*. La Jolla, Calif.: University Associates, 1973, p. 70.

29. LIFE PROCESS:
A STRUCTURED FANTASY

Goals

I. To experience in fantasy the process of life from birth to death.

II. To reflect upon the meaning and values of one's life.

III. To discuss with others the basic existential values of life and their personal as well as corporate meaning.

Group Size

Unlimited.

Time Required

Approximately one and one-half hours.

Materials

I. Record player or tape recorder.

II. Background music on records or tape(s).

III. Blindfold for each participant (optional).

IV. Public address sytem or megaphone (optional).

V. A copy of Worksheets I and II for each participant.

Physical Setting

Freedom of movement is essential for a successful experience with the Life Process fantasy. Ideally, the room selected should be spacious, free from furniture and other obstacles, and carpeted. If obstacles are permanent, they should be pointed out to the group before the start of the simulation experience.

Process

I. The facilitator states the goals and objectives of the activity, gives a brief overview of the procedure, describes its focus, and emphasizes that participation is voluntary. If participants have had little or no experience with structured fantasy, the facilitator may want to spend a few moments introducing them to sensory awareness and the purpose of guided fantasy activities. In doing

This structured experience was submitted by Jan and Myron Chartier.

149

so he may want to lead persons to explore their creative imaginations, their capacities to communicate nonverbally, and their emotional responses.

II. Participants need an opportunity to "warm-up" to the guided fantasy experience. It is important to create a climate of "readiness to respond" before beginning. The relaxation exercises that follow are suitable for this purpose.

> *Exercise One.* Facilitator: "Find a place to sit down. Listen carefully to directions." (He proceeds to read slowly and allows adequate periods of silence to elapse between instructions related to various body units.) "Take time to tune into yourself. Become aware of your body beginning with your toes. Can you feel them as a part of you without moving them? Move them ever so slightly. Now vigorously. Sit quietly . . . Think about your feet, ankles, and legs. Are they tired, restricted, cramped, or energized? How do they make you feel right now? Shift their positions several times, being aware of the sensations you experience . . . Now think of the trunk of your body. Are you aware of your breathing? Are you able to sense the rhythm of your pulsebeat? What about the pressure of the floor upon you? . . . Now explore your fingers, hands, and arms. Begin by moving them slightly. Rotate them with the obvious intent to explore. Sit calmly . . . Think about your head. Are you aware of its weight? Move it freely. Then sit motionless and serenely while you explore your thoughts and feelings. Can you intentionally make your mind blank? Are you aware of your feelings? Can you label them? Are you controlling your emotions or are they controlling you? Can you change them or do you want to? Think about yourself as a whole person. Ask yourself, 'Who am I?' . . . Open your eyes and relax."
>
> *Exercise Two.* "Lie back on the floor. Now imagine that you are a puppet and I am a puppeteer. I will tell you what to do. Respond to my instructions as you imagine a puppet would respond to the strings attached to its body. Now, ever so slowly raise your left hand. Higher, higher, higher. Hold it there for a second, t-h-e-n let it drop. Raise your right hand above your head. Point to the sky. Place it back at your side. Point your feet out. Raise your left leg. Lower it. Now raise your right leg as if to take a step. Relax as if all tension has been removed from your strings."

III. The instructions for the Life Process fantasy are written so that they can be read by the facilitator in a step-by-step sequence. It is recommended that less experienced facilitators read them as printed; more experienced facilitators may choose to adapt them. The instructions should be read slowly. Participants should be given time to create their fantasy experience through nonverbal movement.

Sensitivity to the mood and spirit of the group determines how fast the instructions are read. Some facilitators may choose to put the instructions on file cards for ease of handling.

The instructions for appropriate background music are incorporated with the instructions provided for the facilitator. They follow immediately after the step notations and are enclosed in parentheses.

Before beginning the script, the facilitator directs the participants to close their eyes or put on their blindfolds.

Step One (No music)

"Who am I? The fundamental fact of our own existence and identity frequently leaves us with more unanswered questions than clear answers and definite directions. Although our search for selfhood is an existential quest in the here-and-now, it also confronts us with the impact of an irrevocable past and the uncertainty of an unpredictable future. One question leads to another. 'Who am I?' 'To what extent am I bound to the past?' 'What are my feelings?' 'Why do I behave the way I do?' 'How am I resisting change?' 'How am I changing?' 'Now who am I?' 'What is my life stance; what are my values?'

"To explore ourselves in a new way, we are going to engage in an imaginative re-creation of the entire spectrum of life as we experience it from the perspective of this moment. Together we will experience life as it emerges, expands, grows, matures, and finally as it succumbs to death. As a consequence of our pilgrimage, we may discover what it means to be ourselves in the process of life. We may also discover what we value in life."

Step Two (Serene, tranquil music that gradually grows more active)

"Select a place where there is sufficient space for you to stretch out by yourself. Lie down and allow yourself to become aware of the mood of the music. Let your mind relax." (The facilitator allows enough time for the group to become still. He moves to the next step as he observes the group exhibiting genuine relaxation.)

Step Three (Serene, tranquil music continued)

"You are alone now, but you are neither afraid nor lonely. Indeed, you like it here. You feel secure. Your needs are met. You are protected in a safe place. It is the safest place you have ever known. You seem to be totally at one with your world. Your world seems at one with you. You are your world. Your environment embraces you gently. Your needs are met. Imagine it now! With your hands, reach out and touch the protective walls that surround you. Move your entire body as if to shift your position in a restricted space. What are your feelings here? Can you appreciate the security? Is it possible for you to abandon all fear? Lie quietly again and experience the peace and warmth about you." (Pause)

Step Four (Active, tempestuous music)

"Now feeling is welling up within you. There is a restless feeling inside. You want to move, to explore. Your protected environment feels good and right to you, but you sense it is time to move beyond the walls that surround you. You long to be free. Take time to experience what it is like to be torn between the safest place you have ever known and the strong desire to be free and unconfined. You want both desperately. Deep inside, however, you know that to gain one is to lose the other. So you must choose. What will you do?" (Pause) "Move your body to express your ambivalence." (Pause) "As you contemplate remaining in the safety of your cocoon-like surroundings, somehow you know that your protective enclosure is no longer willing to sustain you. You are being pushed out whether you want to go or not. Now it is impossible to stay. You find yourself moving slowly, but certainly, into an unknown environment. What are your feelings about your choice? What do you want to do?" (Pause)

Step Five (Active, lively, flowing music)

"You have now emerged into a totally new environment. It is vast compared to the confined space you just left. Space seems endless here. You don't feel as safe and secure as you did. Who will help you meet your needs? Who will feed you, cleanse you, and hold you close? As you cannot as yet move toward those about you, they must come to you. But will they come? Will they respond to your needs? You lie motionless, dependent on others.

"As you wait for others to come to you there are discoveries that you make. Gradually you discover your body. Starting with your head, move the various parts of your body until you are convinced that your body is real and working. When you are persuaded of this fact, sit up as if you are sitting up for the very first time. What is it like to know you can control your own body? How does it feel to you to know you can move about in space rather than remain confined in a restricted enclosure? What kinds of freedom do you anticipate?"

Step Six (Active, lively, flowing music continued)

"Slowly, rise to your feet. Test your legs for walking. Begin moving about as if you have never walked before. Take cautious, tentative steps putting first one foot and then the other out ahead of you. Are you uncertain? Do you think you may fall? What urges you to continue trying to walk? How important is it to you to experience the freedom of spatial movement? Keep trying until you are confident of your power to move. Start slowly. Now move more rapidly to acquaint yourself with your environment. Your environment surrounds you with possibilities. Experience the space about you. You are free in space. In what direction will you go? What will you do? What can you learn? Move your hands through space as if you are encountering your new environment for the first time. What do you think about your new surroundings? To what extent can you accept your freedom? Would you like to be protected again? Are you able to act on your freedom? Which do you value more, freedom or dependency?

"As you move about, you discover the potentialities of your surroundings are many. You delight in the joy of some experiences: the taste of a fresh orange, the coolness of an ocean breeze, the attractiveness of objects in your coat pockets, the smell of clover, the sound of someone humming. You avoid experiences which frighten or displease you: the taste of cod-liver oil, the sting of a bumblebee, the sight of a dead animal, the smell of garbage, the sound of thunder. There is so much to learn from your experiences. You hurry from one to another. Sometimes you feel good about yourself and when you do you move with confidence." (Pause) "Other times you feel doubtful, anxious, and afraid. You are uncertain about yourself. You are unsure of your actions. You move randomly and without purpose."

Step Seven (Active, lively, flowing music continued)

"While exploring your environment you may have already encountered a being like yourself. If you haven't, seek one out now. Use your hands to discover who that person is. In return, allow yourself to be discovered by that individual. By touching each other what can you learn? Do you feel a bond linking you together? When you have completed an encounter continue to explore your world. When you meet another person, try to assess who that person is in relationship to you. In what ways is the person like you? In what ways is the person different? Is the person friendly or hostile? Can you trust the other person? How does the person make you feel about yourself by touching you? Are you affirmed or do you doubt yourself? To what extent do you need to be related to the other person or to what degree do you want to be separate and independent? Which do you value more?" (Pause) "Have you found at least one person to whom you could be committed? Did you express love and concern to that individual as you discovered each other? If you could find that person again would you relate in the same way or would you relate differently?"

Step Eight (Ominous, threatening music that grows in tempestuous intensity)

"Stop! You are aware that something is threatening you. You are frightened and need protection. Your environment is growing hostile. A storm is upon you. There is thunder and lightning. You are afraid and lonely. You need other people now. Move quickly to find them. Discover that they also need you. As the storm breaks about you, huddle together giving and receiving protection in the same act. If you have not found a group, search for one where you can feel comfort, support, and warmth as the storm rages about you." (Pause. If a participant has difficulty finding a group, the facilitator should lead him toward one.) "As the storm rages, become aware of the persons close to you and what they mean to you. Can you experience their warmth? Is it possible that you might not survive without them? How do you feel about this state of interdependence? How confident are you that together you will survive the threat? Do you feel comforted by the presence of others? Can you trust the support of togetherness?" (Pause) "Or would you withstand the storm by yourself?"

Step Nine (Quieter but still active music)

"The storm is passing now. The danger is diminishing. It is no longer necessary for you to remain huddled together. You feel the need to continue exploring your world, to exercise your freedom again. Before you go, however, you want to express your thanks to those who have sustained you through the storm. You want them to know that you understand the world better, that you accept the friendly and the hostile parts of life. Because of their presence not only have you survived, you have grown. Let these persons know how grateful you are and how much you love them. Knowing that you may not have this opportunity again, tell them nonverbally how much their presence has meant to you."

Step Ten (Joyful, light music)

"The world is bright again. The sun feels warm after the storm. You are free to live in joyous expectation. Life is good. Find a place where you can experience the warmth and goodness. Let your being be free. Permit your body to express joy for the world and your place in it, for the safety that comes with togetherness, for the love of other persons; indeed, for the goodness of life. Respond in your free space to the beauty, goodness, warmth, and joy in life. Take a deep breath. Reach out with your arms to embrace the fullness of life. Express your joy in having been born. Be glad that you have met other persons and shared life with them. This is a day for rejoicing and giving thanks. Can you do it? Are there barriers that hold you back or obstacles that stand in your way? To what extent can you overcome them?" (Pause)

Step Eleven (Soft, gentle, serene music)

"You can accept life now. You have considerable understanding of yourself, other persons, and the world around you. You have experienced emotional ups and downs. You have grown weary from exerting so much energy in exploration and protection. You are less willing and less capable of responding creatively to the challenges of life. You remember the past and cling to it. Life in many ways seems to stand still. The past has meaning, but sometimes you doubt—even fear—the present. Gradually you realize that it is a time for parting, a time for leaving the surroundings you have explored. It is a time for bidding farewell to the friends you have made. It is time to go. You feel compelled to find appropriate ways of saying 'good-bye' to the friends you love. You want them to know how important they have been in bringing meaning to your life. Express your feelings to each person you meet. Help him to understand your message. Can you understand his? Can you accept the feelings he is communicating to you? How do you feel about saying 'good-bye'? Can you do it?" (Pause)

Step Twelve (Soft, slow, thoughtful music)

"The time has come for you to break off the relationships you have with your friends. Find a place where you can be by yourself. Select a location where you feel comfortable in being alone. Sit down and let this be a time for reflection. Ponder the meaning of your life. What have you been to yourself? To others? What has happened to you in your life? How do you feel about it? What life experiences were most rewarding for you? Were there some periods when you were anxious, frustrated, lonely, or frightened? How do you feel about these periods? What have you truly valued in your life? Can you accept the realization that life has passed and that there are no more opportunities to change and grow? Life has *been* and *is* now what you have chosen to make it. Reflect on the meaning and value of your life. You have learned much and accomplished many goals. Other dreams remain unfulfilled, but time has run out. Your body resists more learning and exploring. You find moving about painful. It is a time to reflect on the value meaning of your life."

Step Thirteen (Tranquil music)

"Now there is a pull for you to withdraw from life. Let your body respond slowly to the pull of gravity. Lie down. Relax the muscles you have used in living. Begin giving up the life you have just created. Sort out your feelings. Can you allow yourself to let go of the life you have lived? What is your greatest relief in knowing that life is almost gone? What is your greatest anxiety or fear? What aspect of life do you have the most difficult time releasing? Is it the freedom to act and move about through space, the power to control, the right to choose, or the relationships with persons you love? What does this say about what you value in life?" (Pause) "Allow yourself to affirm the life you have lived, to perceive it as having been good *to be*. Finally, let it go." (Pause)

"As you relax and lie quietly, contemplate your experiences of being. Embrace as fully as possible the totality of your life. Recall its meaning. Affirm yourself and those who helped you become. Receive your entire life, and then give it up. Letting it fade into nothingness, submit yourself to the power that exists beyond life. Let yourself rest in the ultimate purposes of reality." (Long pause)

Step Fourteen (Tranquil music)

"Have you given up your life? Have you thought about what giving it up has meant to you? When you feel ready to do so, remove your blindfold, open your eyes, and look about. Sit up and tune in to yourself at this moment. Take a short time to sort out your thoughts. Ask yourself, 'What am I? Where have I been? What does my life mean to me? Where is my identity? What do I value?' "

Step Fifteen (No music)

"Join a group of particpants near you and discuss what happened to you and what meaning the entire experience has for you.

I will give you a list of questions on which to focus your discussion if you choose to use them; however, your group is free to explore the meaning of the experience in any way it chooses. Use your time together to learn about the meaning of life and the values you live by."

IV. The facilitator distributes the discussion questions to the participants and encourages them to engage in an open sharing of their experiences and what they learned about their values. The discussion period should not be hurried. The group needs time for their interaction to develop an atmosphere of trust in which deep feelings and values can be disclosed.

V. The facilitator calls the total group together for a short debriefing experience and a summary of what has been learned.

Variations

I. The fantasy may be followed by distributing the Value Rating Sheet and asking group members to complete the items in light of their recent fantasy experience.

II. The facilitator may choose to rewrite parts of the fantasy script to fit the specific needs of the group.

(A) Ask the participants to make themselves comfortable and close their eyes. Tell them to consider their past by fantasizing that they are watching a motion picture of their lives, focusing especially on the ten most important events. Pause for about ten to fifteen minutes.

"Now imagine that you have left the workshop and are walking through the outside world, where suddenly you see the most significant person in your life approaching you. As you draw closer to each other, your relationship with him/her runs through your mind. How does that make you feel? Now focus on a central issue of your relationship. As you meet face to face, what do you want to do? What do you actually do?" (Pause)

"It is time to leave each other. Are you satisfied with the encounter? Have you learned anything about your values from this experience?"

(B) Share and process the fantasy experience.

III. The facilitator may choose to use a different set of "warm-up" exercises or adapt or expand the ones provided.

IV. Rather than using the discussion questions provided, the facilitator might use a variety of other structures, such as open-ended sentences (for dyads or triads), or he might form a circle of the total group and ask each person to share a particularly insightful moment in the fantasy experience.

References

Chartier, M. R. *How to teach with simulation games*. Pasadena, Calif.: Associates in Human Communication, 1974.

Oden, T. C. *The intensive group experience: The new pietism*. Philadelphia: Westminster Press, 1972.

Otto, H. A. *Group methods to actualize human potential: A handbook* (2nd ed.). Beverly Hills, Calif.: The Holistic Press, 1970.

Pfeiffer, J. W., & Jones, J. E. (Eds.). *A handbook of structured experiences for human relations training* (Vol. II). La Jolla, Calif.: University Associates, 1974, pp. 101-112.

WORKSHEET I:
Discussion Questions

1. What did this experience teach you about life, meaning, and values? When did your life begin? How comfortable were you with creating your own life? Were there times you did not want to live? Did you ever resist growth and change?

2. What emotions did you experience as you lived your life? Were they predominantly positive, negative, or mixed? How did you respond to your various emotions? What connection is there between your emotions and your values?

3. Did you have any peak experiences during your lifetime? How did you feel when they were over? What made them positive? Do you cherish these peak experiences?

4. What moments of your life experience were most negative? Why? As your life drew to a close could you embrace the negative experiences as part of life along with the positive?

5. In what aspects was your simulated life like real life? How was it different? Would you change any of your values?

6. How did you feel about your simulated death? Did you learn anything about your feelings concerning your forthcoming real death? In the face of death, how do you value your life; others' lives?

7. How did you feel about relating to other persons during your life? How important were they to you? Is friendship a strong value for you?

8. How effective were you in communicating with the other persons in your life? Did you feel they communicated well with you? Did you learn anything about your own patterns of nonverbal communication as you related with others?

9. What did you learn about trust, support, caring, commitment, and values as you lived out your life? What did you learn about independence? List the five to six most important values in your life.

10. If you had your life to live over again, would you do it differently? How? Why? What values would you change or emphasize to more fully be the person you want to be?

WORKSHEET II: Value Rating Sheet

In light of your experience with the preceding fantasy, respond to the following statements and questions.

I. *Values*

1. Individually, make a list of your most important present values in order of importance.
2. As a group, generate a composite list of these values.
3. Rank order the list (with each individual voting for three values).
4. Using the five most important values from the list, rate the values on the chart below.

    To what degree do you agree with these values as they relate to your life?

	Strongly Agree	Agree	Neutral	Disagree	Strongly Disagree
Value 1					
Value 2					
Value 3					
Value 4					
Value 5					

5. Share information and any learning.

The Value Rating Sheet was submitted by Jeff Enck.

II. *Belief vs. Behavior: Or What Do You Really Believe?*

Change the question in I to:

To what degree does your behavior in your daily interactions with society reflect the values listed by the group?

	Always	Most of the time	Sometimes	Occasionally	Never
Value 1					
Value 2					
Value 3					
Value 4					
Value 5					

1. As a group, discuss the following questions:
   a. To what degree is your behavior inconsistent with your stated values?
   b. If there is an inconsistency, why?
   c. What values are different in the "outside world?"
   d. What values are most difficult to live by in the "outside world?"
   e. What are you going to do (if anything) to change your stated values or your behavior?
2. Process the session and any learning that occurred.

CHAPTER 3

DESIGN CONSIDERATIONS IN VALUE-CLARIFICATION PROGRAMS

One of the primary goals of this book is to provide sufficient understanding of the value-clarification process so that the reader may go on to develop his or her own design strategies and techniques. In Chapters 1 and 2 we saw the basic theory behind value clarification and many examples of value-clarification strategies and exercises. Chapter 2 was designed to serve as a resource that, hopefully, will spark the facilitator to design his or her own strategies to fit the particular needs of a given group or workshop. In the present chapter, the several variables crucial to designing a value-clarification strategy are considered. Among these are: the participants, the particular group dynamics involved, the level of intervention that is desired, a critique of the strategy design, and an evaluation of whether the strategy meets the intended goal(s) of the workshop.

The value-clarification process is most effectively taught and learned within an experiential format and design. Experiential learning, or laboratory education as it is more properly called, refers to a method of teaching and learning that emphasizes (a) the participant's inner experience and external behavior as the primary content; (b) a psychological climate of involvement, trust, and openness; (c) group structures to focus and maximize learning; and (d) the participant's responsibility for his own learning. Laboratory education presupposes the "agricultural" theory of learning: the facilitator, like the gardener, enhances the conditions of growth by providing a nurturing climate and eliminating obstructions—but the actual growing, or learning, is the task of the seedling, or learner.

Laboratory education makes use of a variety of learning activities which can be placed on an involvement continuum (see Figure 3.1). Value-clarification programs typically include some sampling of all

162

these activities, ranging from books, lectures, and films (particularly for theoretical or cognitive content) to instruments, structured experiences, and intensive small groups (for affective learning). The harmonious confluence of these learning activities becomes visible in the program design, and designing an effective program involves the skillful blending of these components. In addition, designing requires that the facilitator be adept in working with small groups: In other words, the value-clarification facilitator must be knowledgeable about the value-clarification process and value-clarification theory, familiar with a variety of experiential learning activities, and equipped with the variety of leadership skills that working with group process requires. It is perhaps unnecessary to state that the facilitator should be a centered, healthy person committed to his own personal growth.

The design of laboratory education events is discussed in detail by Pfeiffer and Jones (1973, 1975), and Cooper and Harrison (1976). Presented here is a focus on those design considerations particularly germane to value clarification.

Low Involvement . . . . .High Involvement

Didactic: Meaning External to Learner							Experiential: Meaning Internal to Learner		
R	L	EL	D	PT	CS	RP	I	SE	IGG
Reading	Lecture	Experiential Lecture	Discussion	Participation Training	Case Study	Role Playing	Instrumentation	Structured Experience	Intensive Growth Group

Figure 3.1.   Involvement Continuum for
Human Relations Training*

*From J. W. Pfeiffer and J. E. Jones, *Reference Guide to Handbooks and Annuals*. La Jolla, Calif.: University Associates, 1975, p. 2.

PRELIMINARY CONTRACTUAL CONSIDERATIONS

For facilitators working within their own systems—a teacher in his own classroom, a minister in his own church—these preliminary considerations are part of the "givens" of the situation and may already be known. For the consultant who has been asked to provide a value-clarification program for a client, specificity regarding the contract will minimize unwelcome surprises and keep communications clear.

Expectations of Client and of Facilitator

In negotiating a contract, it is important for the facilitator to find out what the client wants, what he hopes to achieve, and what he wants to happen for his group. In turn, the facilitator has the opportunity to express the kind of approach he prefers, the conditions under which he likes to work, and the usual impact of what he does. It is crucial that there be clear communication and understanding of what the goal is and what the facilitator will actually do.

A given client may be looking for light entertainment, interesting intellectual discussion of provocative issues, a lecture, an encounter group, problem solving, conflict management, goal setting, leadership training, team or community building, personal and interpersonal growth, or even psychotherapy. Often, the client is unsure of what he wants or needs. Specific contract negotiation enables both facilitator and client to set realistic goals for the program, and to predict probable outcomes. Contract negotiation gives the facilitator the opportunity to state what he can do and what he cannot do and to be honest with the client about it.

Generally speaking, it is from the client that the facilitator learns about the participants with whom he is to work. Once the facilitator has an understanding of what the client wants, the next step is to find out who the participants are.

Knowing the Participants

RELATIONSHIPS

Do the clients form a "stranger" group (participants who come from different places and do not know each other) or an intact group (participants who know one another, as in a club or a group that works together)? Generally, participants feel freer to disclose personal data in a stranger group than in an intact group—this is at least true in the first phase of a group's development. In a stranger group, the participants know that they will not be seeing one another again, whereas, in an intact group, they know they will have to face one another again on Monday morning.

The type of group may influence the design. A facilitator may use more engaging interventions with a stranger group in the first phase of the program. That is, he can assume that participants will not have undue anxiety concerning personal disclosure within the value-clarification experience. However, the transfer of learning from the group to back-home is usually greater in an intact group. In an intact group, what would be a light intervention in a stranger group can be an impactful intervention, especially when the activity focuses on real personal issues.

AGES AND BACKGROUND

What are the ages of the people in the group? All age groups, according to Milton Rokeach (1973), are interested in a world at peace, family security, freedom, and being honest and ambitious. But different age groups prioritize these values and interests differently. Children in early adolescence value equality, true friendship, and being loving and helpful. More than at any other time in life, eleven-year-olds value being cheerful and thirteen-year-olds value being helpful. Seventeen-year-olds most value equality, wisdom, and being courageous, loving, and independent. Persons of college-age value happiness, self-respect, wisdom, a sense of accomplishment, and being broad-minded and independent. People in their twenties value wisdom and being responsible, forgiving, and loving. People in their thirties value self-respect, wisdom, and being responsible and courageous. People in their forties value salvation, self-respect, and being responsible and forgiving. People in their fifties value salvation, self-respect, and being broad-minded. People in their sixties value a comfortable life, national security, and being clean, broad-minded, and forgiving. People in their seventies most value salvation, a comfortable life, and being forgiving and broad-minded. Of course, individuals in groups will have all kinds of values; for a full discussion of this topic, see Rokeach (1973).

Educational level can also be important. Younger people are more familiar with small-group learning than are people over fifty. Elderly people and middle-aged people without a college education may object to the considerable writing that some value-clarification activities involve.

Participant differences in incomes, nationality, sex, occupation, and racial and religious backgrounds have an influence on what the group will be like and what values they will be interested in exploring. In addition, it is helpful to know whether the group has had previous experience with value clarification, encounter groups, etc. No facilitator works equally well with all kinds of groups. The facilitator should be honest with himself regarding his preferences.

SIZE

A simple but strategic piece of information that the facilitator should know in advance is the number of participants in the group. Facilitators vary on the number of people they are comfortable working with; some are willing to work alone with fifteen to twenty persons, others will not work with any size of group without a co-facilitator.

Time and Length of Program

The time frame of value-clarification programs can be continuous (a three-day or week-long workshop, for instance) or distributed (one three-hour session per week for eight weeks). Small-group research indicates that a combined-time program (e.g., a weekend workshop with follow-up sessions) maximizes learning, but in practice, the program time frame is usually built around the schedules of both the facilitator and the participants.

These preliminary contractual considerations describe the detailed prework that a value-clarification program involves. Other considerations—such as costs, facilitator fees, meeting location, and materials—also need to be considered. Careful attention to these considerations benefits both the facilitator and his client.

FACTORS THAT INFLUENCE THE DESIGN OF VALUE-CLARIFICATION PROGRAMS

Value clarification—as a process and as a set of learning activities—appeals to many different kinds of human relations facilitators. Teachers, ministers, leadership trainers, organizational consultants, marriage counselors, and psychotherapists have welcomed value-clarification methods and incorporated them into their work. Value-clarification activities have taken place in many different kinds of settings, and these activities over the past ten years have surfaced a number of factors that influence the design of value-clarification programs. The facilitator's consideration of these factors can increase the effectiveness of the program design.

The Values of Value Clarification

As an approach to learning, value clarification shares the general values of humanistic psychology, such as respect for the individual, the desirability of self-actualizing processes, and a firm belief in the goodness of a democratic character structure. The method of value clarification emphasizes or values a process focus in learning, the freedom of the individual to discover and form his own values, and resistance to the imposition of values by others.

These values attach to the value-clarification *process*. In contrast, "objective" value or "position taking" is *not* part of the *content* of value clarification. In the value-clarification approach, the facilitator presents himself as knowing how to enable persons to discover their own values through exploration of the value process. He does *not* present himself as an expert in objective values, one who would tell another person what his values should be. A tenet of value clarification is that the facilitator must be open about his own value biases but not impose them on others. The facilitator has a right to his own values, and presumably, has used the value process on himself so that he clearly knows his own values and recognizes when he is imposing them on others. It is usually appropriate for a person to attempt to openly influence the values of another; it is inappropriate for a facilitator to impose values in indirect or manipulative ways.

The absence of "objective" value in value clarification has prompted some critics to view the approach as "hedonistic" and "relativistic." This issue is discussed by Kirschenbaum (1976); for the facilitator and the group he leads, the distinction between the value-full process and the value-free content of value clarification must be stressed.

Goals

Value-clarification activities pursue three general objectives or goals: teaching the cognitive aspects of the value-clarification process; promoting personal growth and the involvement of participants in their own growth; and encouraging action steps that implement participants' values.

Teaching the cognitive aspects of the value-clarification process is best done at the program's beginning. The intent of this teaching is to build trust and openness in the group so there will be a climate conducive to personal growth and, eventually, to constructive confrontation of values. In addition to setting the climate, the teaching of the value-clarification process provides participants with a common vocabulary with which to talk about values.

The pursuit of the second goal, involving the participants in their personal growth in values, is usually engaging and stimulating. Most people want to know about themselves and their values, and want to grow as persons. This goal is a necessary part of any session or workshop on value clarification. It serves as a criterion for whether a particular value-clarification activity (especially any adaptation or experimental activity designed by the facilitator) actually helps a person to discover and form his values.

Activities related to this goal heighten interest in value clarification and build cohesiveness in the group. Value-clarification activities usually help a group to reach rapid cohesion as people share what they are discovering about themselves and their values. This second goal is the primary emphasis in any one-session or one-day workshop. Trust, openness, cohesiveness, and the acceptance of the individual must be present before the person or group moves to a confronting, challenging level.

The third goal is to stimulate the person (or group) to act on his values. This goal fulfills the total purpose of value clarification, and it serves as a criterion for whether a particular activity enables a person to clarify his values. "Acting" is part of the definition of a full value. Pursuit of this goal involves a direct challenge to the participants to act on their values once they have fully explored what their value stance is. To move people to act on their values is part of all value-clarification theory and technique. Any course, series, or workshop that does not strive for this goal cannot accurately be called value clarification.

By keeping these three goals in mind while designing and conducting a program, the facilitator avoids debasing value-clarification activities into superficial parlor games.

Appropriate Content

The primary content of value clarification comes from the experience of participants, and it may include attitudes, interests, desires, feelings, beliefs, worries, problems, dreams, life goals, the meanings one gives to life, ethical issues, social problems, and philosophical or moral concerns.

Topics in value clarification can be classified as lively (interesting from an intellectual viewpoint), stimulating (emotionally involving), or vital (topics that move people to action). In a value-clarification program, the facilitator allows participants to determine the content topics as often as possible rather than making his own decision about what is lively, stimulating, or vital. Otherwise, he risks letting his own biases and interests determine what is to be considered and developed as valid topics for value discussions.

Helping a group to choose topics is a matter of some delicacy. The best procedure is to use some introductory activities that help to evolve topics in which the group is interested, e.g., Love List, Coat of Arms, Brainstorming, etc. By diagnosing the needs and concerns of the group members, the facilitator can design activities that best reflect their interests.

Predicting how a given individual may react to a topic is difficult. For example, an activity that shows how some of our values are introjected from parents will generally be lively or stimulating. However, if there is an individual who has just lost a parent, this topic will not only be vital to him but may very well be upsetting. In a given group session focused on drug use, some will find the topic boring, some will learn new information and find the topic lively, some may feel strongly about the subject and find the topic stimulating, and some who have seen or experienced negative results of drug use may find the topic vital. In addition, there may be parents in the group who are emotionally upset by the topic because their children are involved with drugs.

The variables of the situation-person-topic cluster are not always known by the facilitator. Especially in stranger groups, a facilitator must be prepared to handle the unexpected. A facilitator must strive to know what he can about the group and be aware of what topics are currently popular.

If a topic is at least of some interest, value-clarification activities will involve the participants sufficiently to make the topic even more interesting. The Value-Clarification Sheet is one of the best techniques to assess current topics. The facilitator will accumulate a number of value-clarification sheets on various topics, which can be taken to any session or workshop to provide background. He should continually evaluate the appeal of a given topic for a particular group at any given time. A good rule of thumb is to throw away an old value-clarification sheet before it seems outdated and make a new sheet on a more current topic or revise an old one by finding a more current quote or viewpoint on the topic.

Group Process

The value-clarification process is focused primarily on values; its aim is to educate participants about their values and about how to make decisions based on these values (that is, how to act on them). Yet the facilitator must also remain aware of the ongoing interpersonal and group processes in a value-clarification program, even though he typically does not explicitly comment on them, as would the facilitator of an encounter group. (This dictum applies particularly to groups that have no group-dynamic background; in a group in which participants do have a background in dynamics, explicit comments may be made as a means of enhancing the group experience for them.)

The facilitator's awareness of group process will alert him to such phenomena as dependency on the leader, inclusion problems, control or authority issues, and the intimacy that accompanies self-disclosure. Occasionally, the richness of the ongoing process tempts both facilitator and participant to restructure the program into an encounter group; however, this temptation should be resisted since it will diffuse the value focus.

The facilitator's familiarity with the events of group process will provide him with useful cues regarding timing, pacing, level of interventions, and his own impact on the group. Useful discussions of working with group process are provided by Banet (1974, 1976).

Targets and Levels of Interventions

A value-clarification program consists of a series of interventions designed to impact participants in various ways. It is useful for the facilitator to know the *target* of value-clarification intervention (that is, whether a given intervention aims at the participant's thinking processes, emotional reactions, or the total person) as well as the *level* (educational, engaging, or confronting) of the intervention.

TARGETS

1. *Intellect*: A given value-clarification activity may appeal only to the participant's intellect and ways of thinking. This is a necessary and important first step as an introduction to value clarification, but it must quickly be deepened. If only the intellect of the person is touched, what follows is an interesting discussion but one in which there is no action and no real clarification of values. In value clarification, it is important to affect the total person and it is usually easiest to begin with the intellect. In reality, we are not always able to tell what part or parts of the total person will be affected by a given activity, but for practical purposes, an introductory value-clarification activity may build only an intellectual awareness of the valuing process.

2. *Emotions*: Many value-clarification activities evoke emotional responses. If the activity is isolated from the process and only the emotions are affected, the result is a temporary "high" without incorporated, long-term learning. Some individuals who are "turned on" by an emotional approach do not cognitively integrate the learning, nor move to action steps. One of the most constructive benefits of value clarification is that it moves the person toward a congruence of thinking, feeling, and doing.

3. *The Total Person*: The ultimate goal and purpose of value clarification is to involve the total person—body, mind, spirit, and

behavior—in the valuing process. If the participant is not moved to some concrete action, the value-clarification process is incomplete. Acting provides two of the criteria for a full value. It is this emphasis on acting and doing that makes value clarification a useful component in human relations training—training that involves planning, goal setting, problem solving, and implementation.

When the person has fully formed values and repeatedly behaves in accordance with those values, he is growing and fulfilling his potential as a person. In other words, the total person has been confronted and challenged to grow. In this sense, value-clarification methodology is an ongoing, dynamic approach to growth: intellectually, emotionally, volitionally, and spiritually.

LEVELS

An *educational* intervention incorporates a lively topic that builds intellectual awareness and a cognitive understanding of the value-clarification process.

An *engaging* intervention uses a stimulating topic to involve both the intellect and the emotions in such a way that personal growth in value formation occurs.

A *confronting* intervention involves a vital topic that affects the total person in such a way that the person is moved to action and growth through the value clarifying sequence.

Table 3.1 shows the relationship between goals, topics, targets, and value-clarification programs.

Table 3.1. Levels of Intervention in the Value-Clarification Process

Value-Clarification Process	Levels of Intervention	Topic Quality	Goal	Target
Choosing	Educational	Lively	Cognitive Understanding	Intellect
Prizing	Engaging	Stimulating	Personal Growth	Intellect and Emotions
Acting	Confronting	Vital	Action	Total person: Intellect, Emotions, Actions, Spirit

Sequencing

Table 3.1 may be viewed as a guide to the most useful sequence of value-clarification activities. In general, the level of intervention moves from educational to engaging to confronting; the topics move from lively to stimulating to vital. The initial goal of a program is to build the person's cognitive understanding by involving his intellect; the intermediate goal is personal growth involving intellect and emotions; and the final goal is to encourage action, involving the total person.

Examples of sequencing are provided in the design outlines at the conclusion of this chapter.

Evaluation

Six standards for evaluating any value-clarification activity or design are (1) the quality of the topic, (2) the liveliness of the thinking produced, (3) the depth of the stimulation of the emotions, (4) the alternatives presented, (5) the confronting challenge to action, and (6) whether or not some kind of growth occurs. The check list that follows may be used by the facilitator to critique a new design or activity, or to evaluate the results of a program after working with the group. These standards and the use of this check list will help a facilitator improve his skill at making value-clarifying interventions. Facilitators are encouraged to keep notes on their work with various groups in order to improve their style and to refine their techniques.

CRITIQUE CHECK LIST

Topic:

_____ The topic is either lively, stimulating, or vital. It is a topic the participants are interested in and is worth spending time considering. (Circle *lively*, *stimulating*, or *vital* according to your evaluation of the topic.)

_____ The topic is dull and boring. It provides merely intellectual discussion in a humdrum manner.

Thinking:

_____ Thinking will be affected and be catalytic to reflection, consideration, and the choosing of a value.

_____ Thinking will not be affected. The activity is too routine or old hat or demands only a simple response or repetition from memory. No stimulation of thinking.

Emotions:

_____ The experience affects the emotions of the participant; it penetrates deeply into hidden potential and taps the intuitiveness and creativeness of the person.

_____ The experience is bland, conventional, and superficial. It can be handled off the top of the head and thus does not involve or move the person. Or, it is sentimental and overemotional, thus not congruent with the person (or group).

Alternatives:

_____ The design stimulates the exploration of alternatives and their consequences. Value indicators have been consistently elicited.

_____ The activity does not provide sufficient motivation for the person to search for alternatives or to reflect on consequences.

Action:

_____ The activity focuses clearly on the necessity to take action in some way. The participants' behavior will change as a result of this experience. It may well make a difference in the lives of the participants.

_____ The activity does not provide an awareness of the need for action. It is merely busy work that may be interesting but will not affect anything of significance in the participants' lives.

Growth:

_____ The participant will definitely grow as a total person (thoughts, feelings, actions, spirit). The potential of the person will be tapped. Some significant growth will occur as a result of this activity.

_____ The focus of the activity is too intellectual, or it is so sentimental that no growth is tapped. No growth will occur as a result of this superficial experience.

SOME PRACTICAL GUIDELINES

1. Before using a value-clarification activity with a group, take the time to go through it personally. Then, if possible, test it with a small group before using the activity with a large group.

2. Begin by using one value-clarification activity per group. Gradually increase the number with which you are comfortable.

3. With any group, it is important to build a climate of trust. The value-clarification facilitator must be able to establish rapport with his group and create the kind of climate in which people will feel safe to explore their values.

4. For effective sequencing, begin by using the activities that are at the educational level, and then move to the engaging level. Do not rush into using confronting-level interventions until you have had some experience and feel comfortable with the value-clarification process. However, do not unduly put off experimenting with a confronting-level technique.

5. Emphasize a basic ground rule of value clarification: People are to share only what they feel comfortable in sharing. Do not advertise a value-clarification workshop and then turn it into an encounter group.

6. As you assimilate the value-clarification theory and experiment with value-clarification activities, you will become aware of how many subtle impositions of values there are by people and institutions. Be aware of your value biases and do not impose your values on others in a value-clarification workshop. You may have to encourage participants to allow a value-free atmosphere in which participants may freely explore their values.

7. Some people will come expecting you to tell them what their values should be. They will be disappointed and angry when you do not. Patiently explain the process to them and encourage them to explore their value indicators. Check with them later to see how they are responding.

8. Make the time limits clear, so that there is plenty of time for participants to share what they are discovering with each other. Value clarification blends individual work time and group work time; growth comes from both. Many value-clarification activities serve as catalytic instruments to help spark dialogue and interaction among participants.

9. Allow small groups to work on their own without too much facilitating from you. Allow groups to move at their own pace. Some participants will be threatened by what they discover about their values—permit them to intellectualize the threat away.

10. If some participants get into arguments about objective values, steer them back to examining the process of valuing. Encourage them to strive to listen and understand one another and not to moralize to one another.

SAMPLE DESIGNS

These sample designs illustrate a few of the many ways in which value-clarification theory and activities may be blended into a stimulating, involving program. They demonstrate *a* way to design a program, not *the* way; facilitators are encouraged to experiment with different activities, to invent new ones, and, in general, to personalize their experiential design.

AN INTRODUCTORY OPENING SESSION

This three-hour design can be used in a one-session demonstration program or as the first session in a longer workshop or a series of meetings.

Design for an Introductory Session

Brief introduction of staff,
  statement of objectives (Five minutes)

Value Love List (Sixty minutes)

Lecturette on values and value
  indicators (Ten minutes)

Small-group discussion

Coffee break

Value Incomplete Sentences

Value Rankings (Fifteen minutes)

Small-group discussion (Thirty minutes)

Coat of Arms (Thirty minutes)

Questions

Evaluation

Closing

Objectives

I. To introduce participants to the basic concept of value-clarification theory.

II. To have participants experience some educational and engaging levels of the value-clarification process.

III. To stimulate participants so that they will want to know more about their own values and about value clarification.

Materials

Paper, pens, newsprint, copies of attendant worksheets to the techniques.

Process

I. To open the program, the facilitator introduces himself and reviews the objectives. (Five minutes.)

II. He introduces the Value Love List (page 20). For an introductory session, this basic activity and the sharing should be kept to one hour.

III. After the participants have completed the Value Love List, the facilitator discusses the operational definitions of a full value and of value indicators in a ten-minute lecturette.

IV. He divides the participants into small groups to share reactions to the lecturette and the Value Love List.

V. Break. (Ten minutes)

VI. The participants are given five or six Value Incomplete Sentences (page 50) (Fifteen minutes) and four Value Ranking (page 36). (Fifteen minutes)

VII. The participants are divided into small groups to share what they have discovered through the two previous experiences. (Thirty minutes)

VIII. The facilitator uses Coat of Arms (page 53). (Thirty minutes)

IX. To close and to provide an evaluation of the program, the facilitator may briefly explain voting as a value-clarification technique and have the participants vote on the three preceding activities, indicating whether they liked them, whether they found them interesting, and whether they found them helpful. The facilitator closes by answering any questions and by leading participants in a discussion of what the session meant to them.

Variation

In a series of workshops or in a longer workshop, it is useful to end the opening session with a value-brainstorming activity in which the participants come up with a list of values they want to explore during the remainder of the workshop. The subsequent sessions are designed on the basis of the ideas generated during the brainstorming activity.

A ONE-DAY SESSION

In a one-day session, the value-clarification program is basically introductory; only interventions at the educational and engaging level are suggested.

Objectives

I. To introduce participants to the basic concept of value-clarification theory.

II. To have participants experience some educational and engaging levels of the value-clarification process.

III. To stimulate participants so that they will want to know more about their own values and about value clarification.

Process

I. The morning session is identical to the Introductory Opening Session described on page 176.

II. After lunch, the facilitator introduces the Value-Clarification Sheet activity (page 87), adapting it to a value issue in which the group is interested. (Ninety minutes)

III. Coffee break. (Fifteen minutes)

IV. The facilitator conducts the One Value of Friendship activity (page 93). (Thirty minutes)

V. Discussion and evaluation follow. (Twenty minutes)

Variations

I. A group of educators would be interested in Value Voting, Value Ranking, Value Continuum, Value Thought Sheet, Value Autobiography, Value-Clarification Journal, Value Questions, Value Interview, Life Raft, or Value-Indicator Search.

II. A group of facilitators interested in personal growth would prefer Value Ranking, Incomplete Value Sentences, Value Interview, Value-Clarification Sheet, The Value of Friendship, Coat of Arms, Value Role Play, or Fantasy Exploration.

III. In an organizational setting, the following activities would be appropriate selections: Value Interview, Value Brainstorming,

Value Problem Solving, Coat of Arms, Life Raft, Value-Indicator Search, Conversion of Limitations, Values and Conflict, or Negotiation of Expectations.

A THREE-DAY INTRODUCTORY WORKSHOP

This design is intended to expose the participants to as many activities as possible. The sequence starts with introductory activities and goes on to highly confrontive ones.

Objectives

I. To give participants a full introduction to the value-clarification process.

II. To expose the participants to as many value-clarification experiences as possible.

A Three-Day Workshop

	First Day	Second Day	Third Day
Morning:		Value Continuum Value Questions	Conversion of Limitations Value Sheet on content of the group's choice
Afternoon:		Value Indicators: Time/Money Value-Indicator Search	Finish up Value-Clarification Journal Back-home application
Evening:	The Basic Session: Love List Sentences/Ranking Coat of Arms List of Ten Values Value-Clarification Journal	Value of Friendship Informal social recreation	

Process

I. The facilitator begins with the basic Introductory Opening Session (page 176). Toward the end of this session the participants brainstorm a list of ten values, and the facilitator initiates the use of the Value-Clarification Journal for the workshop.

(A) The participants are asked to list ten values they would like to work on during the workshop. Then they rank them from one to ten. The facilitator collects the lists and studies them in order to adapt the activities to the needs and interests of the participants and to sequence them according to value-clarification principles. (Fifteen to twenty minutes)

(B) The Value-Clarification Journal (page 64) is introduced. The participants are allowed ten minutes to begin the journal by recording the reactions and learnings from the first session. The facilitator should remind participants throughout the workshop to make use of their journals. At least one-half hour during the last session should be devoted to finishing up their journals.

II. Value Continuum (page 44) and Value Questions (page 66) are the major activities of the second session.

III. For the third session, Value Indicators (page 123) is used. It is more useful to work with one value indicator rather than several; then the Value-Indicator Search (page 133) is introduced.

IV. The Value of Friendship activity is saved for the fourth session to help increase a strong cohesiveness in the group. This session may be only two hours to allow time for socializing.

V. For the fifth session, the two activities are Conversion of Limitations (page 137) and Value Sheet (pages 58,87), based on the content of the group's selection.

VI. The Value-Clarification Journal is completed. The focus for back-home application may consist of having the participants choose one value that has emerged during the workshop and having them use Value Problem Solving (page 77) to plan how they are going to implement that one value in their lives.

A PROFESSIONAL DEVELOPMENT WORKSHOP FOR GROUP FACILITATORS

This design is for those who have had some experience as facilitators of personal growth groups or laboratory education programs. The design attempts to expose the participants to as many activities as possible. Lecturettes may be interspersed throughout the workshop at the facilitator's discretion.

A Professional Development Workshop

	First Day	Second Day	Third Day
Morning:		Value Indicators: Time/Money Value-Indicator Search	Participants design their own activities Value Continuum Value Sheet Negotiation of Expectations
Afternoon:		Value Problem Solving Conversion of Limitations	Back-home application Finish journal Value Role Play Value Interview
Evening:	The Basic Session: Love List Sentences/Ranking Coat of Arms Value Brainstorming Value-Clarification Journal	Values and Conflict Fantasy Exploration	

Objectives:

I. To introduce participants to the value-clarification process, theory, and practice.

II. To expose the facilitators of groups to many facets of value clarification for personal and professional development.

III. To demonstrate how value clarification may be applied to other group processes.

Process

I. The opening session follows the basic Introductory Opening Session design but adapts the content to the probable interests of the facilitators. About one-half hour is taken to do some brainstorming to discover what values the group is interested in. Another fifteen minutes is taken to introduce the Value-Clarification Journal.

II. For the second session, Value Indicators: Time and Money (page 123) is used to build skills prior to using the Value-Indicator Search (page 133), which incorporates all of the indicators.

III. The third session focuses on Value Problem Solving (page 77) and Conversion of Limitations (page 137).

IV. Values and Conflict (page 144) and Fantasy Exploration (page 139) are used in the fourth session.

V. For the fifth session, participants design their own value-clarification strategies for the morning session of the last day and present them to one another.

VI. In the final session, Value Role Play (page 131) and Value Interview (page 71) are used. Any approach the facilitator wishes is used to plan for back-home applications.

A SERIES OF EVENING VALUE-CLARIFICATION SESSIONS

This design encompasses light evening sessions, each with a different theme.

Objectives

I. To introduce participants to the basic concept of value-clarification theory.

II. To have participants experience some educational and engaging levels of the value-clarification process.

III. To stimulate participants so that they will want to know more about their own values and about value clarification.

Process

For the weekly course in value clarification, appropriate sequencing is important. A usual progression of themes is from introductory material and focus on the individual to a focus on interpersonal relationships and significant life issues.

Each weekly session typically focuses on a particular theme with one or two activities presented in detail. The following themes constitute a suggested format for a series of weekly sessions:

A Series of Evening Sessions

Session	Theme	Activity
I	Introduction	Introductory opening session
II	Identity	Coat of Arms (page 53) Value-Clarification Journal (page 64)
III	Changing Times	Value Sheet (pp. 58,87) Value-Clarification Journal (cont.)
IV	Friendship	Value of Friendship (page 93) Value-Clarification Journal (cont.)
V	Relationships with others	Value Role Play (page 131) Value Indicators: Time/Money (page 123)
VI	Self-awareness	Fantasy Exploration (page 139) Value-Clarification Journal (cont.)
VII	Life Styles	Group designs its own activity
VIII	Sex and Life Styles	Group designs its own activity
		Closing

Concluding Remarks

These sample designs are provided as illustrations. The combinations of activities, theoretical material, and media are limited only by the imagination of the facilitator. From each presentation, the facilitator learns; in each subsequent presentation, the facilitator refines, alters, and perfects what has been done before. Effective designing requires attention not only to the participants and their needs, but also to a consideration of what is happening inside the facilitator.

CHAPTER 4

READINGS IN VALUE CLARIFICATION

The articles that follow are provided to give the value-clarification facilitator an opportunity to enlarge his theoretical perspective of the field through studying and reflecting upon some of the underlying issues of the value-clarification approach. Ultimately, the facilitator must struggle with these issues to arrive at a coherent and integrated understanding of the theory and practice of value clarification. It is hoped, then, that these articles will serve as a springboard for initiating the process of learning and growth that leads to understanding.

Brian Hall, in his paper "Values: Education and Consciousness: The State of the Art, Challenge in our Times," gives his evaluation and perspective of the value-clarification approach as he sees it today. He reflects briefly on the history of several approaches to values and on such primary contemporary theories as character education (traditional moralists), value clarification (Raths, Simon, Harmin), moral development (Kohlberg), and critical consciousness (Freire, Illich). Hall concludes by succinctly outlining his confluent theory of values, which stresses the development of consciousness and the integration of instrumental, interpersonal systems, and imaginal skills of the person in four major phases. The reading provides an overview of current thinking in value clarification.

Arvid Adell is highly critical of the value-clarification approach as he perceives it. In "Values Clarification Revised" he criticizes its weaknesses on the grounds that it is based on two misconceptions. First, he contends that value-clarification theory reifies values as separate entities that exist apart from the persons who hold them; for Adell values are not like vegetables—vegetables have a tangible existence, values do not. In addition, Adell maintains that the value-clarification approach is one-sided in that it concentrates exclusively on the *process* of valuing, thereby ignoring or denying the *content* of

values. Second, Adell maintains that the value-clarification movement
is biased, dogmatic, and moralistic while genuinely believing itself to
be open and pluralistic. He considers the approach to be superficial and
oversimplified, treating only symptoms and not bothering to trace values back to their source. Adell argues that the approach should deal,
during the valuing process, with the depths of the person and that it
must consider the complexity of the many value orientations and
viewpoints that a person holds. Any facilitator who prizes openness
will want to consider Adell's remarks seriously.

One of the most basic developments of value clarification is the
"clarifying response." This skill, or technique, may be compared to the
"third ear" of psychotherapy or to the Rogerian therapeutic practice of
"listening to the feelings within the content of statements." To help the
facilitator develop this value-clarification technique, the chapter titled
"The Clarifying Response" from Raths, Harmin, and Simon's *Values
and Teaching* is reprinted here. This selection discusses the criteria of
an effective clarifying response, gives thirty examples of clarifying
responses, details the uses of the clarifying response in terms of the
seven valuing components, explains what clarifying is *not*, and pro-
poses a method for utilizing the clarifying response.

For years, Milton Rokeach has been prominent in value research,
and included in this section are two selections from his book *The Nature of Human Values*. The first selection, Chapter I of the book, describes Rokeach's interpretation of what values are. He distinguishes
between terminal and instrumental values and presents an instrument he developed for determining what values Americans hold. He
discusses the functions of values and value systems, and distinguishes
values from attitudes, social norms, needs, traits, interests, and orientations. He also discusses the antecedents and consequences of values
and value systems and works toward a classification of values. The
second selection, "Age Differences in Values," discusses the results of
Rokeach's research showing that different values are more or less
prominent at different ages. This selection has been included for its
relevance to the facilitative process.

Carl Rogers, in his article "Toward a Modern Approach to Values:
The Valuing Process in the Mature Person," describes the process
whereby humans acquire values—from parents, religious leaders,
teachers, peer groups—and discusses the outcome of this acquisition in
terms of typical values held by most adults in contemporary Western
societies.

Rogers believes that the values most of us have introjected often
are at variance with our experiences and with our organismic selves.
He sees this "fundamental discrepancy" between our intellectual or

external notion of ourselves and our experiencing or internal notion of ourselves as being at the root of modern man's self-estrangement. Thus, for Rogers, one of the fundamental aims of therapy is to reunite man with himself through procedures leading to personal growth and maturity. In this reading, Rogers outlines the elements and outcomes of his therapeutic valuing process.

One of the oldest controversies among humans revolves around the question of whether values are subjective or objective. Risieri Frondizi explores this controversy in his book *What Is a Value?* For Frondizi, the basic question is "Are things valuable because we desire them, or do we desire them because they are valuable?" In this selection from his book, both sides of the question are presented and the necessary distinctions between them are made in order to better understand this dilemma. Frondizi maintains that the problem will not be resolved by siding either with the subjectivists or the objectivists. He suggests, rather, that values may be the result of a tension between the subject and the object and he therefore offers both a subjective and an objective point of view. A brief hierarchy of values is also presented. In conclusion, Frondizi sets forth another value problem, "How do we apprehend values?" In other words, that people value something does not say that they *should* value it.

A favorite activity of value theorists is to attempt to classify values. Nicholas Rescher presents his viewpoint in the selection "The Dimensions of Values." Rescher explores six main principles for classifying or differentiating between values: (1) their subscribership; (2) their object items; (3) the sort of benefits at issue; (4) the sort of purposes at issue; (5) the relationship between subscriber and beneficiary; and (6) the relationship of the value to other values.

Values are a complex area of human life. Dealing with values almost necessarily leads one into philosophical and theological disputes, and as a result many teachers and/or facilitators shy away from value issues when dealing with their students or clients. Jerrold R. Coombs favors confrontation on value issues and examines what this entails in the selection "Objectives of Value Analysis." Coombs discusses the clarification of the term *value*, the four possible objectives of value analysis for students, the features of evaluative reasoning, the standards of rational value judgments and, within this context, the objectives of value analysis. Part of his conclusion is that there are two defensible objectives of value analysis in the classroom: (1) helping students make the most rational, defensible value judgments they can make, and (2) helping students acquire the capabilities necessary to make rational value decisions and the disposition to do so.

Sooner or later the serious facilitator of value clarification will want to deal with his own philosophy of value. To this end, Wolfgang Köhler presents a penetrating philosophical paper titled "Value and Fact." Köhler deals with such issues as value and the scientist, value and psychology, and how value appears in common experience: the phenomenology of the value situation.

The study of values, like that of psychology, has a long past but a short history. The philosophical dilemmas that preoccupied Aristotle and Plato and the riddles confronted by Lao-tsu and Confucius endure into our century. These readings represent a spectrum of contemporary positions on the process of value formation and clarification.

VALUES: EDUCATION AND CONSCIOUSNESS: THE STATE OF THE ART, CHALLENGE IN OUR TIMES

Brian P. Hall

INTRODUCTION

First let us try to review some of the thinking about value education by putting it into perspective. It seems that everywhere you go today, people are talking about the values we do or don't have. At every conference there is a speaker discussing values. It seems to me that in this society with its Hollywood mentality, things often get distorted and out of perspective. We find value education being examined along with transcendental meditation, parent effectiveness training, and home management courses. We need perspective and a context in which to view these educational issues and methods.

HISTORY IN THE ABSENCE

I was surprised to read in the June, '75 issue of *The Kappan*, which was dedicated to value and moral education, that no historical precedents were given. Current authors such as Kohlberg and Simon were discussed as if they had emerged from a vacuum, as if the issue of values were a brand new thing.[1]

Some historical background will, perhaps, allow us to place present thinking, including my own, into perspective. Having sketched a bit of history, I would like to scrutinize briefly the world view behind the viewpoints of various authors. Then we shall examine some of the questions they are posing to people like us; and indeed, to society in general.

HISTORY PRESENT

No matter what you call it—moral education, value education, or whatever—the issue is as old as civilization. Aristotle talked about a hierarchy of goods, and value educators tend to talk about value ranking or prioritizing significant options in our lives.[2] It seems to me that the process of value clarification has not only been around a long time, but that it is actually a part of what human beings are all about. Albert Camus points out that people judge and oppress

From an Inaugural Address delivered at Seattle University, November 1975, by Brian P. Hall, President of the Center for the Exploration of Values and Meaning. Reprinted with permission from author.

one another because of their values. But he admits that it is intolerable and inhuman to be without them.[3] Western history reflects this tension. After the death of Christ there followed four hundred years of intense discussion as to what *his* value was: Was he a man or a God or fifty-fifty? Then came the Great Councils, which tried to refine the questions and answers, too! Then they put the cork on the bottle by trying to institutionalize some of their conclusions. When Luther nailed his objections to the church door in the sixteenth century, the cork blew off again and we went into another whole period of clarification—called the Reformation.

Value clarification is simply a part of what we're all about; we continually try to describe where we are coming from and to clarify what our values really mean. As Kenneth Clark has pointed out in his classic work, *Civilisation*, it's quite possible that the French Revolution was caused by the writing of the dictionary.[4] Before the writing of this book, politicians were able to use all sorts of value-laden words which no one understood; but when the meanings were clarified by the printing of the dictionary, the language of politics and of the elite became very clear, and revolution ensued.

HISTORY AND PRESENT IDEAS

Let me put that last remark into perspective. I would not want to suggest that value issues, nor other issues for that matter, should be dealt with primarily through verbal or cognitive discussions. The issues arise because of the history—the lived experience and the visual and symbolic interpretation of it—that we are all involved in.[5] Take, for example, the value of liberation of women, When we look at Western art and the images of woman around the tenth and eleventh centuries, we see her as a brutish-looking person, often characterized as an old hag carrying huge rocks to build the cathedrals of Europe. Two hundred years later we see a transformation in art forms and in the cult of the Virgin Mary. Suddenly there are statues of women fragile and delicate in nature. There has been a shift in consciousness of what woman is. This was not particularly a religious shift, although clearly the religious community helped to bring about the new consciousness.

It's no coincidence that during this period people began to place a high priority on relaxation. Some people had more money, and so they could spend more time making love to one another. But these people were primarily from wealthy families, and were often related to the royal courts. And so the value of courtesy or courtly love emerged. It's interesting that even the visions of St. Francis and his conversations with the Lady Poverty were couched in the language of the Court—of courtly love.

Clearly the issue of women's liberation is still with us. It's an international issue now—the result of a growing consciousness and of a continued process of clarification. The issues of population, of birth control, of economic development and of the War are all interwoven with the issue of women's liberation. That is why the United Nations is spending so much time on the subject.

We see then that values education is by no means a new issue or subject matter. What is new is the kind of discussion that is going on now, which strangely enough seems to be centered in public education—although in fact the arena is much wider than that.

VALUE EDUCATION: THE PRESENT DISCUSSION

Let us now review some of the present trends and thinkers. In order to get a grasp of what is going on in value education we must recognize that the major historical question is: Are our values chosen primarily out of a subjective stance or are they primarily objective, out there and given?[6]

The subjectivist argues that each person's experiences and choices are unique. What is valued by one is not valued by another because value is not a quality in things or actions. Rather, things and actions are invested with value because a person takes some positive interest in them or has some good feelings about them.[7] Or the subjectivist believes that each person, by his choices, creates himself, his world, his image of man. Sartre, for example, says each person confers a universal value on his image of man.[8] But the point is that each person individually does the conferring. In other words, what I value in people and in the world, I choose to value, and I alone am responsible for both the choices and for their consequences. This latter emphasis is common to the Existentialist world view.

The objectivist, on the other hand, is likely to find himself struggling to survive in an alien world which is beyond control. He does not choose values, but discovers and cherishes them. He believes that there are certain rules, regulations and laws that one needs to adhere to if civilization is to develop or even to continue. In other words, values are inherent in nature, in social and in personal structures. So values such as honesty, trust, schooling and democracy—the social givens—are to be taught as good for everyone. They are givens which must be passed on to our children.[9]

THE POLES OF THE STRUGGLE

It is not just a value issue in the technical sense which we're about, but the nature of human struggle. We look to the history of ideas and trace man from Aristotle, trying to get truth from objective observations, to Descartes' rationalism spawned British empiricism, which in turn gave birth to logical positivism.[10] And many not-so-sophisticated persons of science have emerged from this history believing that the only truth available to us is that which is obtained by observing measurable data.[11] Measurable data is factual and value-free. It is what *is* and says nothing about what *ought* to be. So science is unrelated to and independent of human values. Philosophically untenable as this view is, the majority of science teachers seem to believe it.[11]

On the other side of the coin, there is a whole movement rooted in the birth of psychoanalysis and described by people like Philip Rieff, which sees man as caught primarily in a struggle with himself.[12] There are the rages of the unconscious, on the one hand, and social norms on the other. One's life ends up fundamentally centered in harmony around himself. The future holds no need for politics or many of the social structures which we enjoy, only individuals harmoniously alone. Paradise is subjective choice and inner peace.

We have characterized objective/subjective stances in an attempt to gain perspective on what is going on in value education. Now let us review some of the modern writers in order to understand where they are coming from.

THE PRESENT SITUATION

Character Education

The most common stance in education today is what I would call moralism or character education. This isn't spelled out as a particular philosophy; but it's the one generally found in school curricula, and particularly in religious education.

The secure but oppressive atmosphere often derives from moralism's objective stance. There are certain rules, certain morals and certain values that the teacher/church/society 'ought' to pass on to people. Its most negative stance is to view children as 'those rather primitive little urchins who need to be controlled and nurtured in their innocence.' They gain virtue by assimilating the values of society that the educational system pours into them. Values are a part of the currency in the educational banking system described by Freire.[13]

Its most sophisticated stance would be the recognition that people do need certain controls, that we do have problems and limitations within our social system, and thus we need laws and given values in order to develop.

However, this is a very objective stance, and the question that is raised by its 'world view' is: What kind of controls do we need to develop a controlled society? It's a law-and-order stance. The limitation here is that it fails to take into consideration developmental educational theory. *This stresses that people do not in fact accumulate values as banks do money, but they internalize and assimilate them only as there are free opportunities to choose from a series of alternatives.*[14] Further, children learn values not so much from what is given verbally, but rather from what is experienced behaviorally.

Value Clarification

Most people in educational circles tend to think of value clarification when they think of Values Education. Further, they often think of Raths, Simon, Harmin and some of the materials that I myself have produced.[15] These materials take a subjective stance. The goal of value clarification is basically to help people clarify or see what specific values are operational in their behavior—to help them look at their own choices by examining their actions.

The main agenda, often hidden, is to show people that what they believe, verbalize and imagine is often quite different from the values that are actually demonstrated in the way they behave. The process of developing a value is described by Raths, Simon and Harmin only in terms of those things which are consciously chosen. Values are not values unless they are chosen consciously from alternatives after consideration of the consequences. They are personally prized, publicly asserted and acted on repeatedly and recently.[16]

The point which Simon is trying to get across, and I feel he does it well, is not so much that his criteria are the last word in a philosophy of values, but rather that the public education system needs to be more sophisticated in the way it tries to get children to assimilate or develop values. The values are in fact developed through free choice; they are developed out of the environment that basically allows the child to choose. The implicit values in a teacher's behavior are more likely to be picked up by the child than the values the teacher talks about. This would be Simon's approach.

When we examine the perspective here and see what its limits are, it doesn't matter who's doing the value clarification or how sophisticated he is; in fact, the method itself comes out of a specific stance. It is rooted in an Existential view which is very subjective in nature. The intended consequence of a value clarification exercise is that you may come to know what 'is,' but never what 'ought to be.' It is strictly amoral. That is not to say it isn't useful, but I think that as a method it is very limited. I find it interesting that many religious educators pick up on the method, use it, and assimilate value clarification exercises into religious education curriculum—without recognizing what the underlying value assumptions of its methods are!

The key questions that value clarification raises are: 1) How is it that we assimilate values? 2) What values have I assimilated? and at a deeper level, 3) What are the values which I am in fact imposing out of my own behavior when I teach, act or behave in a certain manner?

Moral Development and Lawrence Kohlberg

The third strain or movement in the value area is called Moral Development. Its primary exponents are Lawrence Kohlberg, and also the Ontario Institute of Moral Development (OIMD) in Canada.[17] Kohlberg poo-poos what he calls Aristotle's bag of virtues, but he names *justice* as the ultimate value in his Platonic universe. Justice is an *objective* reality to be assimilated and lived out subjectively. With Kohlberg there is in fact much more of a bond between the subjective and the objective than one finds in the existentialist theories of value and morality.

Kohlberg approaches the problem developmentally, harking back to sounds of Dewey, Piaget and Havinghurst. The more underdeveloped the individual, the more subjective and self-centered is his choice. The higher his cognitive development, the more objectively he begins to choose values. Kohlberg, for the most part, is concerned with strictly cognitive, rational structures. I think most people are unaware of how limited his scheme of moral development is, and I believe Kohlberg would be the first to admit it.

His work is a creative extension of the work of Piaget. For Kohlberg the key to the way people choose in regard to what is right and wrong has less to do with data/facts than with the way they interpret the data. And that has to do with where they are in their own moral development. A person could know all the facts about population and starvation in a given country, but in Kohlberg's view this wouldn't make any difference; one would still make choices in terms of which stage of moral development he is in. There are six stages, and I'll just describe them in pairs.

STAGES OF MORAL DEVELOPMENT

The first two stages of moral development are characterized by the person who decides what is right, what he 'ought' to do, based on need for physical gratification. A little child will go into someone's house, see some candy, pick it up and eat it. If you tell him that it is someone else's candy, that it's wrong and he shouldn't do that, he simply will not understand; because for the small child, what is right is to satisfy his own senses. He likes candy, so he takes it. From Kohlberg's point of view, you don't tell him he's wrong; you just pick him up and put him somewhere else. When he matures he will realize it was someone else's property and behave differently.

The second two stages of moral development are characterized by people who decide that what is right is based primarily on what their peer group or external authority thinks is right. As Attorney General Mitchell put it when asked if he thought he did anything wrong: "I put my trust in the role of the President." External authority was the norm here. If you saw the Watergate trial on television you probably noticed that there was almost nil genuine communication; Mitchell just did not understand what they were asking him. What was right for him was external authority. The criteria of internal conscience is available only at a higher stage. For persons in these middle stages, external authority may be the Bible or the Church's authority or the *Sayings of Chairman Mao*. This is not to say that authorities are wrong or bad or unnecessary; it is simply to say that a person at this stage of development will choose what is right on the basis of what external authority says—without criticizing it. It is right because the authority says so. This is not a criticized or evaluated stance of authority. In these stages people shift from a subjective view of morality to an objective one.

The third set of criteria (Kohlberg's stages five and six) are made on the basis of conscience. Stage six people make decisions as to what is right based on law, but law which is criticized and evaluated. For example: "I'm not going to believe in this passage in the Bible unless I feel it's true!" Or: "I'm not going to go along with a new law unless I've thought about it and agree with it."

Stage six moves to the concept of intuitive conscience; the highest level of moral development where a person intuitively knows what is right. For Kohlberg, there is a kind of objective, cross-cultural criteria here for justice and truth.

What Kohlberg has done has great value, and is a significant theoretical offering. His world view attacks a static view of man, and makes us look at everything in terms of development. He attacks religion and also traditional ethical studies, which insist that you should make moral judgments on the basis of knowledge. Moral judgment is based on reason, and reason is the product of development. He raises the question, then: "How can we make clear moral analyses? Who should be judge?" In terms of his view of development, the answer appears to be: "Those few who have reached Stage VI."

However, this theory is found to be wanting because the perspective is strictly in terms of cognitive development. Beyond the exposure to reasoning about moral dilemmas by persons one stage ahead, nothing is said, given or talked about which helps people to move from one moral stage to another! A considerable amount of research has been done in Canada, and very little has happened as a consequence. It is good theoretically, but simply impractical—so far, at least.

This criticism is not so much of Kohlberg as of his followers, who try to utilize the material not recognizing what its limits are or what its world view is.

Critical Consciousness: Freire and Illich

These are persons who have much to say but are rarely heard of in this country, I think because politically they are inaccessible to some people. Possibly they exert a stronger influence than we even know or understand. Generally they have a particular theoretical stance which is directed at the grass roots, the poor and oppressed of this world. They are people like Paulo Freire, Ivan Illich and Everett Reimer. There are many others, but these will serve to illustrate our point. These people speak of the value issue as being the issue of critical consciousness. I can explain this better by sharing a personal experience, since I have worked with Freire's method myself.

AN EXPERIENCE

I was working in a village in Central America which was very heavily deprived economically; the village was in a state of absolute poverty. I was working in a village of 40,000. My Spanish was not very good, and only an Indian language and Spanish were spoken. My experience was to stand on street corners for four of five weeks, trying to listen for what was considered a value-laden word, or "consciousness word." In this particular instance, the word most repeated was "problem," a very simple word. I worked at a church and at a social agency, and went around daily to people in their homes asking, "What is this thing you keep talking about, the *problem?*"

The first thing they said was that the problem had to do with the state of the children of the barrio, malnutrition and the problem of economic prostitution. Prostitution started at around the age of nine years there.

Now as a group of twelve persons we started classes on health, abortion and birth control. People came mainly to see the movies and slides we showed, and nothing happened. They were not interested in discussing what we thought was the issue. So we went back to the drawing board and discussed this problem: What was it?

Well, what happened over a period of four months was that we became more intimate as a group, and got more and more into the issue. The problem came to be described not as economic prostitution but the fact that the women of the barrio wanted more children. This was because having children was related to their sense of "self worth" even though the children starved as a consequence. So the problem did not have to do so much with too many children as with its connection with self worth. As that realization began to dawn on them, the group got together, and with money given by the Costa Rican government, started a Family Education Clinic which was able to deal directly with some of their problems. It was staffed by the very people themselves. By providing those same women with an opportunity to talk to other women about issues of birth control and how this affected them, the clinic became a symbol of possibility. Self-worth comes to be seen as something for everyone and as attainable in many ways—not just through bearing children.

This sort of process is what Freire calls conscientization. The word that we used, just the one word *problem*, was value-laden. Freire has discovered that people learn to read and write very rapidly when value-laden words are chosen as the focus of learning. As Minister of Education for the Brazilian government he had also found that the consciousness of the poor expands critically as to the ills of their society when they become literate. One of the fruits of his successful methodology was that Freire came to be viewed as a political "troublemaker" and was run out of Brazil.

Along with Freire's idea that value issues are the basis of conscientization we also have the analysis of people like Ivan Illich and Everett Reimer, who say that values can not be taught in schools when the system of schooling is itself so value-corrupt. The public school systems, they point out, severely limit the values of students by imposing many of its own institutional values through rigid rules and grading systems. The present educational, political and medical structures are non-convivial in that no one can control or direct them. The recipients of their services are powerless! Illich is furious that a person in American society cannot even choose if he wants to die or live anymore. Machines and doctors take over that responsibility and sick people are subjected to unsolicited and often unwanted treatment. The medical *system* has gotten so out of hand that malpractice insurance and the possibility of legal suits is of more concern to some doctors now than the right to live—or the "right to die."

Freire and Illich discuss the value issue from the point of view of the renewal of society. What they are actually talking about is a new kind of society based on human liberation.

Liberation for Freire lies in conscientization of the little people. Revolutionary change will occur as these people become aware of their own power and their resources for solving their own problems. What is underlying world view here? Illich said to me recently, "In my view man is a suffering servant. Man is the suffering body of Christ. He suffers because of the systems he has created; therefore, I want to expose the systems so that man can renew them and become liberated from them." Illich, in effect, suggests that we should become Mendicant beggars. We should stop supporting non-convivial consumer oriented institutions and wander around the face of the earth being friends to our brothers. So he talks about getting rid of automobiles because they pollute the air, and about slowing traffic to 20 m.p.h. With such changes there would be more intimacy and fewer deaths. Impractical? Maybe, but these theorists have a vision and a world view that many in the international realm feel are our only possibilities for survival.

The Center for the Exploration of Values and Meaning:
Hall, Smith, and Cantin

At CEVAM a confluent theory of values, has been developed which presumes that all these value movements are aspects of one whole, and that the real issue is not values *per se*, but the development of the consciousness itself of individuals, of institutions and of societies.[18] The consciousness of individuals is often objectified in institutions and in societies; so the development of the consciousness of individuals is a crucial element in the development of social consciousness as is social consciousness crucial to the individual's development. Our own consciousness develops in phases, and as it develops certain value options become evident to us. That is to say, the values which we choose are limited by the consciousness that we have. Another way of saying this is that values are consequences of our world view. Raising this question in terms of some of the thinkers we have mentioned, I would like to approach this by looking at the area of skills.

SKILLS

It seems to me that Kohlberg and educators in general are stressing the need for cognitive skills—what we at CEVAM call instrumental skills. Teachers have to teach people *how* to do things; how to read and write; how to think logically when coming to a moral decision; how to shake test tubes in a chemistry lab and how to repair television sets in an electronics course. The focus of instrumental skills is on technique or *how to* whether the emphasis is physical or upon the cognitive structures.

Value clarification and the whole emphasis on affective education reflects the conviction that if you really want to be a well-rounded human being, you are not going to be able to get along with cognitive or instrumental skills only. Unless you have good interpersonal relationships, can relate well to people, and are in touch with your feelings you are liable to wind up leading a robot type of life. The emphasis here is on interpersonal skills, and from this point of view value clarification is basically an interpersonal strategy.

Freire and authors in his camp are raising questions about systems. To us there is an issue of system skills. It is not just individual behavior. The emphasis on "person" in this country, which is largely a result of the psychological movements of the last ten years, has much to do with the phase of development that society is in—middle class American society that is. We have found that many problems that appear to be personal or individual are in fact system problems. As individuals are pulled in all directions by the system within which they operate, they find they have no system skills with which to cope. Now the issue here is that any given system (or institution) relays its own values to a person and reinforces certain values in that person. As a consequence, the values imposed or reinforced exclude or limit other values that might be chosen and acted on.

For example, in the last social agency with which I consulted, I was struck by the fact that the majority of counselors were social workers trained in a particular school of social work. The major emphasis in the agency was on counseling in a one-to-one relationship. From the agency's perspective, healing a person required that clients actually came into the office to talk out their problems.

This procedure raised the question for me of what happened to a client's self esteem as he came down the street and walked into the agency—let us call it "The Goodwill Social Services." Without having been seen, this client was asked to accept something about himself which might not be appropriate or true.

This taken-for-granted perspective on healing and service raised other questions for me and for some of my colleagues. We found the majority of our clients were teenagers. It soon became evident that all they really needed was a bit of tender loving care; so it was suggested that the best thing to do for them would be to take them somewhere—to the Art Museum perhaps—and to just be with them for an afternoon. The resistance to this was incredible. The social work supervisor claimed, "You can't do that; social work process has to do with one-to-one counseling. Furthermore, we are not at all sure we believe in all these wild new therapeutic techniques that are emerging!"

I am not exaggerating; this really happened! What I am trying to translate to you is that in reality, it would not matter who the client was or how open the counselors were, the value of "healing" had been defined in such a way as to "limit" the ways in which healing the client could occur. This was a value that came down not from individuals but through the system.

Clearly, it would not matter how open a given counselor was nor what kinds of values he had, he would be limited by the values of the system. Simply stated: The way in which we set up structures (for example, at a university) or the way we choose members of the board (for example, on the basis of power or on the basis of personal creativity) limits or expands the values of that institution. The way we set up structures in our institutions determines how values will precipitate down through and limit the people within the structures. The skills of structuring creatively and thus of enhancing human development are called system skills.

Public education curricula need to take system skills as seriously as it takes interpersonal and cognitive learning skills if education is to be holistic. The child in the classroom should know what the teacher is teaching him by her behavior—the method by which she is teaching. Educators need to look at the world view that is operating in the classroom through its structure.

A few weeks ago we were asked if we would work with some white teachers who were having problems dealing with black students being bused to their school. The issue was: "Can you come and help these teachers understand the values of those black children?

We said, "No, we can't do that. But what we can do is help the teacher to understand what values she's operating with, what values are being imposed upon her by the administrative system, what values are being imposed upon that administration by the local public . . . and what values the kids are ending up with!" She may not be able to change anything, but at least she will know the name of the game—and she might be less uptight as a consequence. Our underlying conviction, of course, was that conscientization does make a difference.

The purpose of learning and of teaching system skills is not to bring about the overthrow of systems. I think one of the most naive things about the tremendous emphasis on individualism in this country is that people do not understand that they are operating in systems. People always move faster than systems; and since systems are always going to be slower than people, good leadership is always going to be deviant—it is always going to be ahead of where systems and where most people are! Knowing this may not change the system, but it may help to prevent an ulcer or the breakup of an administrator's family.

A consciousness approach to problems raises the question: "What should be the minimal level of consciousness for leadership in any institution?" The response is important since leaders will be reinforcing all sorts of values through their institutional systems. This reinforcement could come from a parent in a family, a teacher in a classroom of from the president of a university. Whether the system is large or small, intimate or remote, the impact of the leaders' own consciousness on the experiences and consciousness of others in their system is real and often pervasive.

IMAGINAL SKILLS

The fourth skill, without which the other three are impossible, is imaginal skill. This is probably the thing most lacking in educational systems. I mean by imaginal skills, the ability to see where we are going, and what our view of the future is, in a value-specific way. I mean *vision*.

The question of morality is not just a matter of how we reason about ethical choices, but also a question of integrated skills. For example, former President Nixon—no matter what your political views are on the Watergate scandal—had a pretty good vision of where he wanted to go. He had terrific systems skills and he had good instrumental skills, but his interpersonal skills were zero. Our point is that moral thinking and moral integrity disintegrated because he lacked a holistic, integrated skill base. What we are specifically raising through value education, then, is the relationship between holistic, integrated skill development and the ability to make ethical choices.

PHASES OF CONSCIOUSNESS

Our efforts at CEVAM have been concentrated on the development of a theory of consciousness and values which integrates many other developmental theories such as those of Piaget, Jung and Erickson. For us this meant recognizing that consciousness itself is developmental and that certain core values become dominant depending upon one's phase of development. We posit eight stages of development in four phases of consciousness. The first phase of consciousness—whether it be of an adult or a child—results from viewing the world as hostile and capricious. Obviously values such as self-preservation and security will predominate if a person lacks skill or the environment lacks the resources necessary for supporting life. You can not talk to a person about property investments or ultimate human development if he is hungry!

The second phase of consciousness, often characteristic of a middle class society, results from viewing—the world and life as a problem to be coped with. The central value that emerges in this phase is self-worth. There is an emphasis on success, achievement, duty, belonging and being appreciated, and on values such as order and prestige.

In this phase values which might be quite different are often confused. For example, the value of work is often confused with the value of self-worth. A person behaves as if his credo were: "I am what I do!" Or, "I am what I have worked for!" Media advertising reinforces that: "You owe it to yourself to buy a Cadillac—you are really worth it." It is a sign that you have achieved in this world; it doesn't matter whether or not you can pay for it. It's image and prestige that count in this second phase.

In the third phase values that stress independence, rights, liberation, and justice emerge. Here the world view is one of life as a call to creative invention: "I will take charge." This phase parallels Kohlberg's fifth and sixth stages where conscience and internal authority emerge. Individual choice and honesty are supreme. It is at this phase *the people are* likely to rush off to encounter groups and to feel the need to share *who he is* with everyone else.

In the fourth phase the world view goes beyond Western-style independence to high levels of cooperative existence. In this world view people see themselves as life-givers. The key value moves from independence to interdependence. Now the values of intimacy and solitude and conviviality rise to priority. Tools, technology, and institutions are limited so their control is possible by every man. In this view, medical practice is no more esoteric or exclusive than arithmetic practice; both are appropriate to general education. Cities are totally planned and slums eliminated. Administrations and leadership cooperate intimately eliminating one-man shows.

As the phases of development expand, they do so not only in terms of personal development, but also in terms of world view. The earliest stage in my world is simply the physical me and the extensions of me in my family and home. Later on, the perspective expands to the local village, to the nation and ultimately to a global vision.

CONCLUSION

In attempting to outline what is called a confluent theory of values we have stressed the importance of a holistic approach—one that is cross-cultural and cross-disciplinary. Values are seen as the consequence of one's experience, skills and world view. This presentation is itself limited. When you come down to it, it is also the expression of a particular world view! So we disappear within ourselves no matter how hard we try to be clever, intelligent and objective. We are becoming more aware as we develop our theory how really limited and imposing, as a matter of fact, our own world view is.

Now, I think that the difference between talking about world view in terms of value education and world view as we traditionally talk about it in terms of theology, philosophy, etc., is that we are trying to make it behaviorly specific to people. I believe that in the future persons who are teaching value education will have to be willing to look at various world views and also to expose their world views to other people. We can see what the world views are that are bombarding us in society and that we are getting through television. It appears to me that one of the goals of a Christian educator, or of any educator for that matter, is to provide students with skills in value discernment and to help them avoid the picking up of whatever values are passed along the supermarket counter, rather than to promote willy-nilly. Educators and in turn students should be conscious of what their own world view is, so that they can choose and make a stand when views alien to theirs are being imposed. Professors/teachers should not only expose students to their own world views, but also give guidance as to what world views may be destructive or negative to reinforcing or developing the individual and society creatively.

In conclusion, when we speak of value education we must ask what it has to offer and why it is important for us to learn about it.

1. Value education is an integrative, holistic discipline: it integrates science, religion and the social sciences. If it has any place as subject matter, it belongs in late high school and college. It should then be seen as an opportunity for students to integrate and bring together all the things they are learning, and to translate them in terms of their own human behavior.

2. Value education expands critical consciousness. The task of education is to expand people's consciousness. This means dealing with a student's behavior and world view and not just dealing with subject matter—whether it be English literature, chemistry or whatever—from the point of cognitive input or particular skills.

3. Value education develops personal authority and maturity. Whatever stance you take or whatever value issue you are addressing, when placed in the context of valuing, you are turning the choice over to the individual. You have to make the individual responsible for his choices; it matters not whether he is choosing from an objective or subjective stance. The expansion of consciousness has to do with the development of responsibility. Therefore, educational methods should always be pushing a person forward in terms of his capacity to make choices.

4. Value education is raising the question of the relationship between moral choice and integrated skills. It is raising this question particularly through some of the political events that have occurred. It is conceivable that a person has no choice but to make what we normatively consider immoral choices if he doesn't have sufficient skills integrated in his person.

5. Value education challenges not just the values of an individual, but it also is beginning to put this in perspective with the institutional environments in which we live. We have found it fairly easy to talk about

values in individuals; and we have found it pretty easy to discuss values of the society or culture in which we live; but I don't hear many people having the courage to talk about values being reinforced through the institutions to which they belong. It is not easy for us either; we get overwhelmed by the systems that we are part of and that are a part of us, as you do.

It takes courage to look at the system as a whole, to ask for the budget, to look at the administrative procedures, and to figure out what values are being imposed or reinforced or enhanced in an individual by the institution. This is not to say that those values are negative; I am simply saying that we should check out what they are, and not take things willy-nilly. Otherwise, we may end up with things we would rather not have.

NOTES

1. *Phi Delta Kappan*, Volume LVI, Number 10, June, 1975.
2. Aristotle, *Nichomachean Ethics*, Book I, Chapter 13. Raths, Louis E., Merrill Harmin, and Sidney Simon. *Values and Teaching*. Columbus, Ohio: Merrill, 1966 and Postman, Neil, and Chas. Weingartner, *Teaching is a Subversive Activity*. New York: Dell, 1969.
3. Camus, Albert. *The Plague*. New York: Random House, 1947.
4. Clark, Kenneth. *Civilization*. New York: Harper and Row, 1950.
5. Janson, H. W. and Dora Janson. *Picture History of Painting from Cave Painting to Modern Times*. New York: Abrams, 1957.
6. For a thorough discussion of this question and its philosophical strains see *What is Value?* by Risseri Forondizi. LaSalle, Illinois: Opencourt, 1971.
7. Perry, Ralph Barton. *General Theory of Value* (Second edition; Cambridge, Mass: Harvard University Press, 1950) pp. 115-116. See also A. J. Ayers *Language, Truth and Logic*. New York: Dover Publications, 1950.
8. In trying to escape the charge of subjectivisms Sartre insists: "In fact, in creating the man that we want to be, there is not a single one of our acts which does not at the same time create an image of man as we think he ought to be. To choose to be this or that is to affirm at the same time the value of what we choose, because we can never choose evil. We always choose the good, and nothing can be good for us without being good for all." The latter statement comes as close to essentialism as Sartre will ever get. But his insistence that existence precedes leaves him in the subjectivist camp. Though he argues that "a man who lies and makes excuses for himself by saying 'not everybody does that,' is someone with an uneasy conscience, because the act of lying implies that a universal value is conferred upon the lie," it is obvious that it is the person who chooses to lie who *confers* the value. Sartre, Jean-Paul. *Existentialism and the Human Emotion*. New York, Philosophical Library. pp. 17-19.
9. The objectivist would be likely to utilize *inculcation* as the most appropriate means of passing on the givens and the values he ascribes to. For a description of this method see: *Values Education Sourcebook*, ERIC, Clearinghouse for Social Studies, Social Science Education, 1976.
10. *Encyclopedia of Philosophy*. Editor-in-chief, Paul Edwards, 8 volumes, New York: MacMillan, 1967.
11. Ayer, A. J. *Language, Truth and Logic*. New York: Dover, 1950.
12. Ruff, Philip. *Freud: Mind of the Moralist*. Garden City, New York: Doubleday, 1961.
13. Freire, Paulo. *Pedagogy of the Oppressed*. Translated by Myra Ramos. New York: Herder & Herder, 1968.

14. Choosing from among alternatives is one of the components of the valuing process outlined by Raths, Harmin, and Simon, op. cit.

15. Hall, Brian P. *Value Clarification as Learning Process*. Three volumes. New York: Paulist Press, 1973.

16. These are the seven criteria for distinguishing genuine values from aspirations or value indicators. Raths, op. cit.

17. Beck, C. M., B. S. Crittendon, and E. V. Sullivan. *Moral Education*. Toronto: University of Toronto Press, 1971. See also *Phi Delta Kappan*, Volume LVI, June, 1975, for several articles by these same authors.

18. For a full-blown discussion of the relationship of consciousness to value development see my book; *Development of Consciousness: A Confluent Theory of Values*. New York: Paulist Press, 1975.

VALUES CLARIFICATION REVISED

Arvid W. Adell

Recently a neighbor asked me to assist in a diagnosis of his unhealthy lawn. Armed with an encyclopedia of herbs and rushing in where only agronomists should tread, we quickly set about to effect a cure. After trying several futile remedies, we finally consulted an expert who informed us that the source of the ailment was to be found not in the grass but in the soil. We had been treating the symptoms rather than the cause.

There is something analogous taking place in education today, particularly in the teaching of values. Educators are being importuned to assist students in identifying and clarifying their values. The urgency with which the task is approached implies that something is amiss: that students are confused about what they prize and hold sacred, and that the product of such confusion is an unproductive and unrewarding life. A group of educators who have dedicated themselves to doing "values clarification" have published several books, including a manual containing 79 values games and quizzes (*Values Clarification: A Handbook of Practical Strategies for Teachers and Students*, by Sidney B. Simon, Leland Howe and Howard Kirschenbaum [Hart, 1972]). No doubt the book will remain popular, but, like the movement itself, it is based on two misconceptions which will inevitably lead to a confusion rather than a clarification of values.

The first of these misconceptions is the error of reifying and hypostasizing values—treating them as though they were independent entities existing apart from persons. Although explicitly the proponents of values clarification seek to avoid this error, implicitly they allow it to infiltrate and sabotage their writings. The second misconception stems from a lack of self-understanding and self-perception. The values clarification movement views itself as genuinely open and pluralistic when in fact it is biased, dogmatic and moralistic. Both of these misconceptions result from the oversimplification of the program: instead of tracing values to their source and analyzing the cause of the malady, values clarification deals with symptoms and appearances. Where it needs to penetrate deeply into the soil, it treats only the grass.

VALUES AND VEGETABLES

What is a value? Although the movement's purpose is the clarification of values, a working definition of a value is lacking. This omission is intentional. The program does not concentrate its energies upon the *content* of values, but rather upon the *process* of evaluating. Presumably there are at least two ingredients in evaluating: the values (the *content*), and the *process* whereby values are selected and arranged hierarchically. This bifurcation of evaluation into content and method is misleading and eventuates in a reification of values. The following example will serve to illustrate how this confusion comes about.

Observe a man deliberating at the frozen-foods section of the grocery. Assuming that he has a vested interest in the item he is selecting, that he prizes and cherishes it, that he has no mandate from home, that there are at least two alternatives, that he dares to affirm his choice publicly and act upon it, we might agree that his process of selection is commendable—at least from a values-clarification point of view. Although his selection of cauliflower rather than lima beans might seem trivial and, in our judgment, bad aesthetics, we laud him for his process-form and we conclude that an evaluatory process has taken place.

The confusion of the values clarification program is its implication that choosing a value is like choosing a vegetable: there are a variety of values to be affirmed, and so long as a person chooses freely, deliberately and purposively, he or she is performing well. But values are *not* like vegetables: they are not objects, entities, things-in-themselves, nouns. Instead, they are indicators, appearances, symptoms of something deeper and more substantial.

If a student states his values and arranges them in some kind of hierarchy, it is not as though he has organized a list of vegetables in order of preference. Nor is the confusion due to the consideration that values are more complicated and subtle than frozen foods. The mistake is that vegetables tangibly exist and values do not! If someone states his values, he isn't giving you a statement of fact about something we call "values." He is telling you something about himself, where he is coming from, what he wants to do, and why he wants to do it.

Therefore, to ask a student to participate in a game or quiz in which he encounters a number of values and arranges them according to his inclinations, and then to rest assured that the task if complete, is to substitute a beginning for a conclusion. To reify a value confuses rather than clarifies because it suggests that values are items which can be isolated, diagnosed and then treated, if necessary—and such is not the case.

If values are not objects but indicators, what is it that they point to? They describe the person doing the self-evaluation, and they inform us of the way in which he views reality. Values unveil a person's metaphysics; they tell us what he has come to believe is important, and if the process continues far enough, they manifest *why* he thinks and behaves the way he does. Since the values-clarification enterprise neglects to trace values back to the life-understandings which give rise to those values, it remains a peripheral operation dealing with shadows and bogus entities.

VALUE-NEUTRAL OR VALUE-LADEN?

The values-clarification program is incisive in its refusal to define and deal with the *content* of specific values because values have no content. The only subject matter which remains for analysis is the *process* of evaluation, and it is here that the second misconception is located. Values clarification purports to be value-neutral; it seeks to offer a process of selection and elucidation that will be applicable to all persons, regardless of their moral attitudes and histories. Thus, the process itself must be objective, disinterested and value-free; only if these conditions are met can it claim to be sufficiently flexible and unbiased to serve a pluralistic society.

Is the process value-neutral and universally applicable? No—and this flaw can be demonstrated by observing how the technique works in a hypothetical situation. Suppose that a group of students is trying to determine the value of premarital sexual experience as opposed to premarital chastity. One of the students advocates premarital experience and demonstrates how this value can be affirmed and defended by using the values-clarification process: i.e., it is highly prized, is chosen freely from alternatives, is rewarding, and adds zest to living.

Another student, who affirms chastity because his religious faith proscribes any kind of premarital or extramarital sexual involvement, discovers that the values-clarification process is incompatible with the kind of moral reasoning he has used. He has not chosen chastity freely from several competing alternatives. His is an authoritarian ethic which makes sense only within the context and framework of his religious understanding of life. Ironically, to affirm "chosen freely" as one of the canons of positive valuation in this situation would, for this student, be immoral. "Choosing freely" is not a value-neutral activity and, as an essential part of the values-clarification process, it illustrates that the process itself is implicitly value-laden.

At this point, the teacher of the values-clarification method is in a dilemma. Either he must depart from the prescribed process of valuation and accept both of these seemingly contradictory positions as valid, thereby exhibiting a genuinely pluralistic attitude toward values, or he must insist that the authoritarian approach is unacceptable a priori because it violates the contention that all values must be freely chosen in order to be authentic values.

To admit that the values-clarification process has presuppositions and biases is not to make a pejorative judgment. That process did not come into being *creatio ex nihilo*, and it has a history dating back to the conceptual framework of John Dewey. Dewey, a naturalist and an empiricist, viewed values, morals and ethical judgments as instruments or tools devised by human beings for dealing with the environment. Thus, he saw valuation as a pragmatic concern, noting that societies value those activities and items which promote growth and bring satisfaction. The subprocesses of values clarification are elaborations of this Deweyan understanding of life: in order for an activity or thing to be classified as valuable (as fostering growth and satisfaction), it must be freely chosen, personally desired, capable of repetition, publicly demonstrable.

But if a person does not share this understanding, he or she may not necessarily share the Deweyan theory of values and valuation either. Obviously, not all persons are empiricists and naturalists, nor are all of us instrumentalists about ethics, nor can we all employ a values-clarification process derived from naturalism and instrumentalism. The process may appear value-neutral to those persons who find themselves in agreement with that view of reality, but to those who do not, it will be viewed as slanted, dogmatic and intolerant.

A GENUINE PLURALISM

If values are indicators pointing beyond themselves to a particular view of life, and if there is a variety of values indicative of a plurality of life-pictures and life styles, it follows that the clarification of values necessitates an illumination of these various understandings. Those who would engage in values clarification must trace values back to their source, clarify that source, and compare it with other ways of viewing the world. Only in this way can the movement assist students in self-understanding rather than subtly inculcating them with its own values.

Admittedly, this process complicates the task considerably, for there appears to be an innumerable variety of life styles and world views. But philosophers who have examined various methodologies of exploring reality and of tracing values have offered several classifications into which most students will fit. Although the following list is not exhaustive, it is probably inclusive enough for most students to be able to relate their thinking to one or more of them.

1. *An empirical approach:* Empiricism asserts that all of our knowledge—factual and valuational—originates in sense experience of our environment. There are no innate, a priori values which antedate experience; nor are there any extrasensory perceptions or other transcendental or esoteric sources of genuine knowledge. By use of observation and experimentation we can achieve and measure results which can then be translated into good and bad, valuable or worthless.

To illustrate this method at work, consider the argument of the students as to the value of chastity or premarital sexual experience. The propriety of choosing either of these alternatives would depend on the obtainable benefits. If experience has shown that premarital involvement fosters personal growth and relationships, then this activity is warranted. On the other hand, if it can be empirically verified that abstinence contributes more to the overall happiness and growth of persons, then this alternative ought to be followed.

2. *The authoritarian approach:* Authoritarianism presupposes that values are not democratic decisions. Certain powerful leaders—societal, parental or religious—have established standards for others to follow. Most persons are regarded as incapable of determining their own values. This incapacity may be due to a lack of experience, talent or status. Presumably, the authority who does determine value transcends at least some of the major limitations which the devotees possess.

One who subscribes to authoritarianism would contend that a person is not free to experiment in certain areas and that measurable results are not the criteria for determining the propriety of an act. If sexual freedom is proscribed by the appropriate authority, this mandate must be obeyed.

3. *A rationalist approach:* Recent studies at the Harvard Center for Moral Education conclude that the structural development of a person's moral consciousness is related to the development of one's cognitive faculties. As a child becomes increasingly mature and aware, he passes through ascending levels of moral sensitivity. The final and highest stage is the rational one, in which a person discovers universal, abstract, apodictic principles whereby he makes moral judgments. These principles transcend any specific moral rules, but moral acts can be judged in the light of them; they are not dependent on measurable results as in empiricism, nor on extrinsic command as in authoritarianism. Instead, they are a priori and universal obligations which are obvious to individuals who have progressed to this level of rational thinking.

In the chastity-premarital-sex debate, the rationalist would seek to find the compatability of this act with a rational definition of value. The relevant questions are these: Can I universalize, in principle, this act? Am I deciding on the basis of independent, abstract, autonomous thinking, or am I following the dictates of feelings or persuasions? Is this act conformable to my interest in justice and obligation to everyone? The irrelevant questions are the ones proposed by the empiricists and the authoritarians: Will this bring more good than bad results? What will others think?

4. *An atheistic existential approach:* Some persons are not convinced that the world exhibits good logical form. For them, there are no objective, demonstrable, a priori truths discoverable by the mature mind. Nor does the world manifest any intrinsic values. The honest, authentic individual is the person who acknowledges the disturbing fact that all value and all truth are subjective, that each person must play god, create all values, and seek to interpolate them into existence.

In our case study, the existentialist would assert that there is no way of measuring value scientifically, nor are there any transcendental authorities who can decide for others, nor are there any objective, universal rational principles of "oughtness" which correlate with the way things are. Each person must decide for himself or herself, and the basis for this decision is the person's will or volition. Like all other value judgments, sexual experience or chastity is a subjective matter, and individuals must create and substantiate their own values. One's only responsibility is to oneself.

5. *An intuitive approach:* Occasionally persons will contend that a particular act is either valuable or worthless and yet be unable to explain why this is the case or how they have arrived at this certitude. Their certainty is not based on measurement of results, external authority, rational thinking or personal decision. The veracity of the judgment is self-evident, immediately perceivable, prereflective and indisputable. Moral judgments are felt to originate from a moral sense—a sixth sense which immediately perceives goodness and badness in the same way that the five senses perceive colors, sounds, tastes, odors

and textures. If a person lacks this moral sense, he is deemed amoral; if he disobeys it, he is acting immorally.

Intuitionists admit that their view is problematic, especially in reference to particular values in ambiguous and complex situations. However, they insist that the basis of all morality and valuation is founded upon a self-evident, nondemonstrable intuition; viz., that a person *ought* to do the good and choose the valuable rather than desire the bad and choose the worthless. If a person lacks this fundamental insight, there are no other possible grounds for moral dialogue.

An intuitionist in favor of chastity would argue that it is self-evident that sexual experience divorced from permanent personal commitment is unconscionable, and that all morally sensitive persons know this.

BEYOND GAMES AND QUIZZES

Obviously this proposed revision of the values-clarification program is considerably more demanding and complex than the original. It would require students to go beyond the games, the quizzes, and the application of the values-clarification process and to become involved in an analysis of themselves and the kinds of presuppositions which form the foundation for their decisions. It would demand a willingness to speak openly about the subterranean centers from which their values surface, and it would require the courage to examine and defend this center. It would probably necessitate reading and discussing thinkers who have articulated these various understandings.

Nevertheless the options seem clear: either the program can continue to describe values as things-in-themselves and deal with them in a superficial manner, or it can become genuinely pluralistic by probing beneath values to the persons who evoke them. Only by choosing the latter alternative can the movement accomplish its stated objectives of assisting young people to see themselves as unique and valuable persons in a complex world.

THE CLARIFYING RESPONSE

Louis E. Raths, Merrill Harmin, and Sidney B. Simon

The basic strategy of this approach to value clarifying rests on a specific method of responding to things a student says or does. This basic responding technique is discussed in this chapter.

Fundamentally, the responding strategy is a way of responding to a student that results in his considering what he has chosen, what he prizes, and/or what he is doing. It stimulates him to clarify his thinking and behavior and thus to clarify his values; it encourages him to think about them.

Imagine a student on the way out of class who says, "Miss Jones, I'm going to Washington, D.C., this weekend with my family." How might a teacher respond? Perhaps, "That's nice," or "Have a good time!"

Neither of those responses is likely to stimulate clarifying thought on the part of the student. Consider a teacher responding in a different way, for example: "Going to Washington, are you? Are you glad you're going?" To sense the clarifying power in that response, imagine the student saying, "No, come to think of it, I'm not glad I'm going. I'd rather play in the Little League game." If the teacher were to say nothing else at this point other than perhaps "Well, we'll see you Monday," or some noncommittal equivalent, one might say that the student would be a little more aware of his life; in this case, his doing things that he is not happy about doing. This is not a very big step, and it might be no step at all, but it might contribute to his considering a bit more seriously how much of his life he should involve in things that he does not prize or cherish. We say it is a step toward value clarity.

Or note this example. A student says that he is planning to go to college after high school. A teacher who replies, "Good for you," or "Which college?", or "Well, I hope you make it," is probably going to serve purposes other than value clarity. But were the teacher to respond, "Have you considered any alternatives?", the goal of value clarity may well be advanced. The "alternatives" response is likely to stimulate thinking about the issue and, if he decides to go to college, that decision is likely to be closer to a value than it was before. It may contribute a little toward moving a college student from the position of going because "it's the thing to do" to going because he wants to get something out of it.

Here are two other samples of exchanges using clarifying responses.

STUDENT: I believe that all men are created equal.

TEACHER: What do you mean by that?

STUDENT: I guess I mean that all people are equally good and none should have advantages over others.

TEACHER: Does this idea suggest that some changes need to be made in our world, even in this school and this town?

STUDENT: Oh, lots of them. Want me to name some?

TEACHER: No, we have to get back to our spelling lesson, but I was just wondering if you were working on any of those changes, actually trying to bring them about.

STUDENT: Not yet, but I may soon.

TEACHER: I see. Now, back to the spelling list. . . .

TEACHER: Bruce, don't you want to go outside and play on the playground?

STUDENT: I dono. I suppose so.

TEACHER: Is there something that you would rather do?

STUDENT: I dono. Nothing much.

TEACHER: You don't seem much to care, Bruce. Is that right?

STUDENT: I suppose so.

TEACHER: And mostly anything we do will be all right with you?

STUDENT: I suppose so. Well, not anything, I guess.

TEACHER: Well, Bruce, we had better go out to the playground now with the others. You let me know sometime if you think of something you would like to do.

The reader may already sense some criteria of an effective clarifying response, that is, a response that encourages someone to look at his life and his ideas and to think about them. These are among the essential elements.

1. The clarifying response avoids moralizing, criticizing, giving values, or evaluating. The adult excludes all hints of "good" or "right" or "acceptable," or their opposites, in such responses.

2. It puts the responsibility on the student to look at his behavior or his ideas and to think and decide for himself what it is *he* wants.

3. A clarifying response also entertains the possibility that the student will *not* look or decide or think. It is permissive and stimulating, but not insistent.

4. It does not try to do big things with its small comments. It works more at stimulating thought relative to what a person does or says. It aims at setting a mood. Each clarifying response is only one of many; the effect is cumulative.

5. Clarifying responses are not used for interview purposes. The goal is not to obtain data, but for the student to clarify his ideas and life if he wants to do so.

6. It is usually not an extended discussion. The idea is for the student to think, and he usually does that best alone, without the temptation to justify his thoughts to an adult. Therefore a teacher will be advised to carry on only two or three rounds of dialogue and then offer to break off the conversation with some noncommittal but honest phrase, such as "Nice talking to you," or "I see what you mean better now," or "Got to get to my next class," or "Let's talk about this another time, shall we?", or "Very interesting. Thanks." (Of course, there is no reason why a student who desires to talk more should be turned aside, the teacher's time permitting.)

7. Clarifying responses are often for individuals. A topic in which John might need clarification may be of no immediate interest to Mary. An issue that is of general concern, of course, may warrant a general clarifying response, say to the whole class, but even here the *individual* must ultimately do the reflecting for himself. Values are personal things. The teacher often responds to one individual, although others may be listening.

8. The teacher doesn't respond to everything everyone says or does in a classroom. There are other responsibilities he has.

9. Clarifying responses operate in situations in which there are no "right" answers, such as in situations involving feelings, attitudes, beliefs, or purposes. They are *not* appropriate for drawing a student toward a predetermined answer. They are *not* questions to which the teacher has an answer already in mind.

10. Clarifying responses are not mechanical things that carefully follow a formula. They must be used creatively and with insight, but with their purpose in mind: when a response helps a student to clarify his thinking or behavior, it is considered effective.

The ten conditions listed above are very difficult to fulfill for the teacher who has not practiced them. The tendency to use responses to students for the purpose of molding students' thinking is very well established in most of our minds. The idea that a function of a teacher is to help the child clarify some of the confusion and ambiguity already in his head is an unfamiliar one for many of us. After all, most of us became teachers because we wanted to *teach* somebody something. Most of us are all too ready to sell our intellectual wares. The clarifying strategy requires a different orientation; not that of adding to the child's ideas but rather one of stimulating him to clarify the ideas he already has.

Here is another classroom incident illustrating a teacher using clarifying responses, in this case to help a student see that free, thoughtful choices can be made. The situation is a classroom discussion in which a boy has just made it clear that he is a liberal in his political viewpoints.

TEACHER: You say, Glenn, that you are a liberal in political matters?
GLENN: Yes, I am.
TEACHER: Where did your ideas come from?
GLENN: Well, my parents I guess, mostly.
TEACHER: Are you familiar with other positions?
GLENN: Well, sort of.
TEACHER: I see, Glenn. Now, class, getting back to the homework for today . . .
    (returning to the general lesson).

Here is another actual situation. In this incident the clarifying response prods the student to clarify his thinking and to examine his behavior to see if it is consistent with his ideas. It is between lessons and a student has just told a teacher that science is his favorite subject.

TEACHER: What exactly do you like about science?
STUDENT: Specifically? Let me see. Gosh, I'm not sure. I guess I just like it in general.

TEACHER: Do you do anything outside of school to have fun with science?
STUDENT: No, not really.
TEACHER: Thank you, Jim. I must get back to work now.

Notice the brevity of the exchanges. Sometimes we call these exchanges "one-legged conferences" because they often take place while a teacher is on one leg, pausing briefly on his way elsewhere. An extended series of probes might give the student the feeling that he was being cross-examined and might make him defensive. Besides it would give him too much to think about. The idea is, without moralizing, to raise a few questions, leave them hanging in the air, and then move on. The student to whom the questions are addressed, and other students who might overhear, may well ponder the questions later, in a lull in the day or in the quiet moments before falling asleep. Gentle prods, but the effect is to stimulate a student who is ready for it to choose, prize, and act in ways outlined by the value theory. And, as the research ... demonstrates, these one-legged conferences add up to make large differences in some students' lives.

THIRTY CLARIFYING RESPONSES

There are several responses that teachers who have worked with the clarifying approach have found very useful. A list of some of these is presented below. As the reader goes through the list, he might make note of some he would like to try; that is, make his own list. There are too many noted here to keep in mind at one time. It is probably best then, to gather a dozen or so together, ones which sound as if they could be used comfortably, and try them out, perhaps expanding or revising the list as experience dictates.

Be reminded, however, that the responses listed here are recommended as useful clarifying responses only when they are used in accordance with the ten conditions listed earlier. The acid test for any response is whether or not it results in a person reflecting on what he has said or done, clarifying, getting to know himself better, examining his choices, considering what he prizes, looking at patterns in his life, and so on. If the response makes the student defensive, or gets him to say what the adult wants him to say, or gives him the feeling that the adult is nagging at him, it is being used improperly or with poor timing. An accepting, noncommittal attitude on the part of the person making responses is crucial.

The reader might note that some of the responses listed below are geared directly to one or another of the seven valuing components: prizing, searching for alternatives, thinking critically, choosing freely, incorporating choices into behavior, examining patterns of living, and affirming choices. Some other responses stimulate reflection in a more general sense. But, in *all* cases, responses are open-ended—they lead the student to no specific value. No one must deliver a "right" answer to a clarifying response. Each student must be permitted to react in his own personal and individual way.

1. Is this something that you prize?

To respond in a way that gets the student to consider whether he prizes or cherishes something he has said or done helps him to clarify his values. The response could, of course, be in a different form and have the same intent, e.g., "Are you proud of that?", "Is that something that is very important to you?", "Is that idea very dear to you; do you really cherish it?" The particular situation in which the response is being made, as well as the age of the child to whom it is directed, will help determine the precise wording.

2. Are you glad about that?

This encourages the student to see whether things he feels, says, or does are things that he is happy about and make him feel good. One could also ask if the student is unhappy about something. Such questions stimulate a child to evaluate his life and to consider changing it if *he* finds it does not bring him satisfactions. Note how different the effect of this response is from the scolding "Aren't you ashamed of that?" Clarifying responses are accepting and illuminating, not rejecting and moralizing.

3. How did you feel when that happened?

It advances clarification for a person to understand that his feelings are part of his understandings and awareness and that they have to be considered in decision-making. He needs to know that feelings are important, that we respect his right to have his own feelings, and that feelings do not have to be suppressed.

4. Did you consider any alternatives?

Note how this tends to widen, to open up the thinking of children (and adults). With this response, as with *all* the others in this list, teachers will need to accept whatever the student replies without judgment. After he answers the question, leave him with an honest "Oh. Now I see," or "I understand," or "You stated your views clearly," or "I appreciate hearing what you say," or some non-judgmental phrase or gesture.

5. Have you felt this way for a long time?

Questions that get at the same thing are, "When did you first begin to believe in that idea?" and "How have your ideas or understandings changed since the time you first considered this notion?" Here the person is pushed to examine the history of his beliefs or attitudes, to look at their origins, and to see if they are really his or if they have been absorbed unthinkingly. Note how the next response might follow after a student replies to this one.

6. Was that something that you yourself selected or chose?

This reminds persons that they *can* make their own choices, if they want to do so. An affirmative reply to this response might well be followed by response Number 7.

7. Did you have to choose that; was it a free choice?

Here, no matter what the student says, it is probably wise to say no more but to discontinue the conversation with some non-judgmental closing.

8. Do you do anything about that idea?

This response helps persons see the responsibility for incorporating choices into actual living. A verbalization that is not lived has little import and is certainly not a value. Another way of saying the same thing: "How does that idea affect your daily life?", or "In what ways do you act upon it?"

9. Can you give me some examples of that idea?

This helps push generalizations and vague statements of belief toward clarity. Note also the relevance of the next response.

10. What do you mean by——: can you define that word?

This also pushes understanding to clarity and helps prevent the mouthing of words students cannot really mean because they do not really understand them.

11. Where would that idea lead; what would be its consequences?

This encourages the student to study carefully the consequences of ideas. No meaningful choice can be made unless the consequences of alternatives are understood. Therefore, it is often very useful to help children examine the consequences of each available alternative. Accordingly, one could also ask, "What would be the results of each of the alternatives?", or "How would those ideas work out in practice?"

12. Would you really do that or are you just talking?

Again the encouragement to see the importance of living in accordance with one's choice.

13. Are you saying that . . . [repeat]?

It is sometimes useful merely to repeat what the student has just said. This has the effect of reflecting his ideas and prompting him to ask himself if he really meant that. It is surprising how many persons seldom hear what they say. Sometimes the phrase "Did I hear you correctly?" can be used for this purpose.

14. Did you say that . . . [repeat in some distorted way]?

Sometimes a teacher does well to purposely twist what a student has said. Will the student attempt to correct the distortion? After trying it, one senses that the effect is much the same as response Number 13.

15. Have you thought much about that idea (or behavior)?

Of course one accepts whatever reply a student makes to this. It is destructive to the valuing process to attack a negative answer to this question with something like, "Well, in the future it would be wise to think before you speak (or act)." An accepting and non-judgmental mood is vital for the valuing process.

16. What are some good things about that notion?

A simple request for justification of expressed ideas in some such non-judgmental words often brings dramatic re-evaluation of thinking on the part of students. Many persons rarely realize that there could or should be good, desirable, worthwhile aspects of ideas they hold. The ideas are just there, unexamined and unevaluated.

17. What do we have to assume for things to work out that way?

Many persons have neglected to examine the assumptions upon which rest their ideas, aspirations, and activities. This probing helps persons understand better, make choices more wisely, and make valuing more possible. It is sometimes useful, in this context, to suggest an assumption that the student seems to be making and ask him if he has considered it, e.g., "Are you assuming that there was *nothing* good about the depression?"

18. Is what you express consistent with . . . [note something else the person said or did that may point to an inconsistency]?

To present such a disconcerting challenge, to note an exception, to relate things with other things, can produce real clarification *if* it is not done with an "I think I have trapped you in an error" tone of voice. The idea is not to slap students down, but to open things up for them so that they can think with new insight, if they want to do so. (Happily, teachers trying this approach seem to find that most students do want to do so.)

19. What other possibilities are there?

This raises alternatives to students and thus it aids them in valuing. Sometimes this question is posed to a group and all alternatives are listed on the board, without judgment again. Of course, other students and the teacher, too, can say which alternative *they* prefer, but there is no judging a child because he chooses a different alternative. No teasing or otherwise deriding others' choices is tolerated or else there is no free choice.

20. Is that a personal preference or do you think most people should believe that?

To inquire whether a statement is intended as a personal preference or whether it is something that should be generally endorsed is one way of helping to distinguish an attitude or prejudice from a social principle. "Is this idea so good that everyone should go along with it?" is another way to get at this.

21. How can I help you do something about your idea? What seems to be the difficulty?

This question reminds the student that *action* is a component of life and intentions are incomplete until acted upon. Sometimes such questions uncover suppressed feelings or misunderstandings. Obviously, they locate real or imagined obstacles, too. Also try: "Where are you stuck?", or "What is holding you up?" (But be prepared to offer help if it's asked for.)

22. *Is there a purpose back of this activity?*

Asking students what, if anything, they are trying to accomplish, where they are headed with ideas or activities, sometimes brings the realization to students—and for the first time—that they might really have purposes and goals and that they might relate their ongoing activities to those purposes and goals.

23. *Is that very important to you?*

This gets students to consider more seriously what is and what is not important to them. It is also often useful to ask students to put several things in order of rank. Assigning priorities is a variation, and a useful one.

24. *Do you do this often?*

"Is there any pattern to your life that incorporates this idea or activity?", one might inquire. The idea here is to help students see what is repeated in their lives and what is not and to leave them with the decision of whether or not to build a pattern.

25. *Would you like to tell others about your idea?*

Inviting a student to explain his ideas to the class or others provides two challenges. It tests to see whether he is committed to his beliefs strongly enough to affirm them in public. It also puts him in the position of thinking through his ideas well enough to explain them, and perhaps justify them, to others.

26. *Do you have any reasons for (saying or doing) that?*

This tests whether or not a choice has been made and to what extent that choice was based on understanding. DANGER: Avoid using that question to pull up short on a student who is obviously not thinking. If you want to tell a student that you believe that he is not thinking, tell him so. But use the above question when you really want to have a student consider his beliefs or actions.

Incidentally, when a student does (or says) something and the teacher inquires, "Sonny, why did you do that?", the student often hears, "Sonny, now why in the world did you ever do something as foolish as that?" "Why" questions are usually to be avoided when attempting to help students clarify their values. "Why" questions tend to make a student defensive, tend to prod him into making up reasons or excuses when he really has none in mind. Besides, the question "Why did you do that?" carries with it the assumption that the student *knows* why, and that is perhaps the reason he tends to concoct a reason when he has none. It is much more effective, for value clarifying purposes, to ask "Do you *have* a reason?" and then sometimes follow up an affirmative reply with, "Would you mind telling me?"

27. *Would you do the same thing over again?*

This helps a student to evaluate things that he has done, to consider why he has done them, and perhaps to affirm the wisdom of doing it in the future. Do not use this question everytime someone does something that *you* do not like. That would be an example of not-so-subtle moralizing. Use the question when you want to stimulate thinking, and strive to keep it non-judgmental.

28. How do you know it's right?

When a child makes a moral or ethical judgment about something by saying that a thing is right or lovely or good, it is useful to ask how he knows that that judgment is correct. Sometimes we ask how he was able to decide. Note this dialogue.

> TEACHER: "I see you're hard at work on that project, Jimmy."
> STUDENT: "It's not good to be lazy, you know."
> TEACHER: "How do you know it's not good?"
> STUDENT: "Everybody knows that. My parents always say it."
> TEACHER: (Walking away) "I see."

Thus may a teacher subtly and persistently suggest that one might think about such matters as rightness, or beauty, or goodness if one wants to do so.

29. Do you value that?

Merely picking out something a student has said or done and asking "Is that something that you value?" helps to stimulate clarifying thinking. Perhaps such a question could have been added by Jimmy's teacher in the above dialogue, e.g.,

> TEACHER: "I see. Is working something that you value then, Jimmy?"
> STUDENT: "Huh? I suppose so."
> TEACHER: "O.K., Jimmy. Thank you."

30. Do you think people will always believe that? Or, "Would Chinese peasants and African hunters also believe that?" Or, "Did people long ago believe that?"

Such questions are useful to suggest to a student that his beliefs may be unknowingly influenced by his surroundings, by his social milieu. It helps him gauge the extent to which he may be conforming. See also response Number 5.

Note Chart 1 for examples of how some of the above clarifying responses, and others, are related to the seven components of the valuing process. Those seven criteria are helpful for thinking of other useful clarifying responses and for keeping in mind the ones above. All clarifying responses in one way or another encourage the student to choose, prize, or act in terms outlined by the value theory.

CHART 1

Clarifying Responses Suggested by the Seven Valuing Processes

1. Choosing freely
   a. Where do you suppose you first got that idea?
   b. How long have you felt that way?

    c. What would people say if you weren't to do what you say you must do?

    d. Are you getting help from anyone? Do you need more help? Can I help?

    e. Are you the only one in your crowd who feels this way?

    f. What do your parents want you to be?

    g. Is there any rebellion in your choice?

    h. How many years will you give to it? What will you do if you're not good enough?

    i. Do you think the idea of having thousands of people cheering when you come out on the field has anything to do with your choice?

2. Choosing from alternatives

    a. What else did you consider before you picked this?

    b. How long did you look around before you decided?

    c. Was it a hard decision? What went into the final decision? Who helped? Do you need any further help?

    d. Did you consider another possible alternative?

    e. Are there some reasons behind your choice?

    f. What choices did you reject before you settled on your present idea or action?

    g. What's really good about this choice which makes it stand out from the other possibilities?

3. Choosing thoughtfully and reflectively

    a. What would be the consequences of each alternative available?

    b. Have you thought about this very much? How did your thinking go?

    c. Is this what I understand you to say . . . [interpret his statement]?

    d. Are you implying that . . . [distort his statement to see if he is clear enough to correct the distortion]?

    e. What assumptions are involved in your choice. Let's examine them.

    f. Define the terms you use. Give me an example of the kind of job you can get without a high-school diploma.

    g. Now if you do this, what will happen to that . . . ?

    h. Is what you say consistent with what you said earlier?

    i. Just what is good about this choice?

    j. Where will it lead?

    k. For whom are you doing this?

    l. With these other choices, rank them in order of significance.

    m. What will you have to do? What are your first steps? Second steps?

    n. Whom else did you talk to?

    o. Have you really weighed it fully?

4. Prizing and cherishing

    a. Are you glad you feel that way?

    b. How long have you wanted it?

    c. What good is it? What purpose does it serve? Why is it important to you?

    d. Should everyone do it your way?

    e. Is it something you really prize?

    f. In what way would life be different without it?

5. Affirming
   a. Would you tell the class the way you feel some time?
   b. Would you be willing to sign a petition supporting that idea?
   c. Are you saying that you believe . . . [repeat the idea]?
   d. You don't mean to say that you believe . . . [repeat the idea]?
   e. Should a person who believes the way you do speak out?
   f. Do people know that you believe that way or that you do that thing?
   g. Are you willing to stand up and be counted for that?

6. Acting upon choices
   a. I hear what you are for; now, is there anything you can do about it? Can I help?
   b. What are your first steps, second steps, etc.?
   c. Are you willing to put some of your money behind this idea?
   d. Have you examined the consequences of your act?
   e. Are there any organizations set up for the same purposes? Will you join?
   f. Have you done much reading on the topic? Who has influenced you?
   g. Have you made any plans to do more than you already have done?
   h. Would you want other people to know you feel this way? What if they disagree with you?
   i. Where will this lead you? How far are you willing to go?
   j. How has it already affected your life? How will it affect it in the future?

7. Repeating
   a. Have you felt this way for some time?
   b. Have you done anything already? Do you do this often?
   c. What are your plans for doing more of it?
   d. Should you get other people interested and involved?
   e. Has it been worth the time and money?
   f. Are there some other things you can do which are like it?
   g. How long do you think you will continue?
   h. What did you *not* do when you went to do that? Was that o.k.?
   i. How did you decide which had priority?
   j. Did you run into any difficulty?
   k. Will you do it again?

SUMMARY

Basic to the use of the approach to values of this book is the clarifying response. The clarifying response is usually aimed at one student at a time, often in brief, informal conversations held in class, in hallways, on the playground, or any place else where the teacher comes in contact with a student who does or says something to trigger such a response.

Especially ripe for clarifying responses are such things as expressions of student attitudes, aspirations, purposes, interests, and activities. These sometimes indicate a value or a potential value and thus we refer to them as value indicators. Also in this category are expressions of student feelings, beliefs,

convictions, worries, and opinions. A teacher who is sensitive to these expressions finds many occasions for useful clarifying responses.

The purpose of the clarifying response is to raise questions in the mind of
the student, to prod him gently to examine his life, his actions, and his ideas,
with the expectation that some will want to use this prodding as an opportunity to clarify their understandings, purposes, feelings, aspirations, attitudes,
beliefs, and so on.

Undergoing this, some students may find the thoughtful consistency between words and deeds that characterizes values. But not everything need be a
value. Beliefs, problems, attitudes, and all the rest are part of life too, although
in most of our lives they might be even clearer and more consistent one to the
other.

It may be useful to list some things that a clarifying response is *not*.

1. Clarifying is not therapy.

2. Clarifying is not used on students with serious emotional problems.

3. Clarifying is not a single one-shot effort, but depends on a program
consistently applied over a period of time.

4. Clarifying avoids moralizing, preaching, indoctrinating, inculcating, or dogmatizing.

5. Clarifying is not an interview, nor is it done in a formal manner.

6. Clarifying is not meant to replace the teacher's other educational
functions.

If clarifying is none of the above, what is it? It is an honest attempt to help
a student look at his life and to encourage him to think about it, and to think
about it in an atmosphere in which positive acceptance exists. No eyebrows are
raised. When a student reveals something before the whole class, he must be
protected from snickers from other class members. An environment where
searching is highly regarded is essential.

We emphasize that students will probably not enter the perplexing process
of clarifying values for themselves if they perceive that the teacher does not
respect them. If trust is not communicated—and the senses of students for such
matters can be mystifyingly keen—the student may well play the game, pretending to clarify and think and choose and prize, while being as unaffected as
by a tiresome morality lecture. This is a difficult and important point, for it is
not easy to be certain that one is communicating trust, whether or not one
believes he is doing so. (A moot point is whether some persons can communicate a trust that they, in fact, do not have.) One must be chary about concluding that a teacher who says the right words is getting the results desired.
There is a spirit, a mood, required that we cannot satisfactorily describe or
measure except to say that it seems related to a basic and honest respect for
students. It may be fair to say that a teacher who does not communicate this
quality will probably obtain only partial results.

For many teachers a mild revolution in their classroom methodology will
be demanded if they are to do very much with the clarification of values. For
one thing, they will have to do much less talking and listen that much more,

and they will have to ask different kinds of questions from the ones they have asked in past years. Teachers usually favor questions that have answers that can be scaled from "right" to "wrong." No such scoring can be applied to answers to clarifying questions.

The rewards for giving up the old patterns may not come right away, but there is mounting evidence that teachers who act "responsively" begin to have small miracles happening in their classrooms. They often see attendance go up, grades rise, and interest and excitement in learning crackle. They witness students who had been classified as apathetic, listless, and indifferent begin to change. In the words of one teacher, "students get their heads off their elbows and use those elbows to wave hands in the air."

In brief, one might see the clarifying response as fitting into the value clarifying method in the following framework:

1. First, look and listen for value indicators, statements or actions which suggest that there could be a value issue involved.

It is usually wise to pay special attention to students who seem to have particularly unclear values. Note especially children who seem to be very apathetic, or indecisive, or who seem to be very flighty, or who drift from here to there without much reason. Note, also, children who overconform, or who are very inconsistent, or who play-act much of the time.

2. Secondly, keep in mind the goal: children who have clear, personal values. The goal, therefore, requires opportunities for children to use the processes of (a) choosing freely, (b) choosing from alternatives, (c) choosing thoughtfully, (d) prizing and cherishing, (e) affirming, (f) acting upon choices, and (g) examining patterns of living. One does this with the expectation that the results of these processes are better understandings of what one stands for and believes in and more intelligent living.

3. Thirdly, one responds to a value indicator with a clarifying question or comment. This response is designed to help the student use one or more of the seven valuing processes listed above. For example, if you guess that a child doesn't give much consideration to what is important to *him*, you might try a clarifying response that gets at prizing and cherishing. Or the form of the value indicator may suggest the form of the clarifying response. For example, a thoughtless choice suggests responses that get at choosing, and a fine-sounding verbalization suggests responses that get at incorporating choices into behavior.

THE NATURE OF HUMAN VALUES
AND VALUE SYSTEMS

Milton Rokeach

Any conception of the nature of human values, if it is to be scientifically fruitful, should satisfy at least certain criteria. It should be intuitively appealing yet capable of operational definition. It should clearly distinguish the value concept from other concepts with which it might be confused—such concepts as attitude, social norm, and need—and yet it should be systematically related to such concepts. It should avoid circular terms that are themselves undefined, such terms as "ought," "should," or "conceptions of the desirable." It should, moreover, represent a value-free approach to the study of human values; that is, an approach that would enable independent investigators to replicate reliably one another's empirical findings and conclusions despite differences in values.

In this introductory chapter a conception of human values will be formulated with criteria such as these in mind. These formulations were guided by five assumptions about the nature of human values: (1) the total number of values that a person possesses is relatively small; (2) all men everywhere possess the same values to different degrees; (3) values are organized into value systems; (4) the antecedents of human values can be traced to culture, society and its institutions, and personality; (5) the consequences of human values will be manifested in virtually all phenomena that social scientists might consider worth investigating and understanding. These assumptions also represent a set of reasons for arguing that the value concept, more than any other, should occupy a central position across all the social sciences—sociology, anthropology, psychology, psychiatry, political science, education, economics, and history. More than any other concept, it is an intervening variable that shows promise of being able to unify the apparently diverse interests of all the sciences concerned with human behavior. "Problems of values," Robin Williams writes, "appear in all fields of the social sciences, and value elements are potentially important as variables to be analyzed in all major areas of investigation" (1968, p. 286).

The value concept has been employed in two distinctively different ways in human discourse. We will often say that a person "has a value" but also that an object "has value." These two usages, which have been explicitly recognized by writers from various disciplines—writers such as Charles Morris in philosophy (1956), Brewster Smith in psychology (1969), and Robin Williams in sociology (1968)—require from us at the outset a decision whether a systematic study of values will turn out to be more fruitful if it focuses on the values that people are said to have or on the values that objects are said to have.

Such a decision is not easy to make because a reading of the literature reveals important writings issuing forth from both camps (Handy, 1970). Perry (1954), Lewis (1962), Hilliard (1950), Thomas and Znaniecki (1918-20), Katz and Stotland (1959), Jones and Gerard (1967), and Campbell (1963), for example, have approached the problem of values from the object side. These writers merely conceive of all objects as having a one-dimensional property of value (or valence) ranging from positive to negative, and the value concept thus has no additional properties or surplus meaning. Even more recently, behaviorist B. F. Skinner has vigorously denied that men possess values and has instead argued that "the reinforcing effects of things are the province of behavioral science, which, to the extent that it is concerned with operant reinforcement, is a science of values" (1971, p. 104). On the person side are such approaches as those of Allport, Vernon, and Lindzey (1960), Kluckhohn (1951), Kluckhohn and Strodtbeck (1961), Maslow (1959, 1964), Charles Morris (1956), Brewster Smith (1969), Robin Williams (1968), and Woodruff and DiVesta (1948). Are there theoretical grounds for deciding which approach is likely to be the more fruitful?

Robin Williams, who has explicitly raised the same question, has remarked that a person's values serve as "the *criteria*, or standards in terms of which evaluations are made. . . . Value-as-criterion is usually the more important usage for purposes of social scientific analysis" (1968, p. 283). One implication that follows from Williams' observation is that the number of values-as-criteria that a person possesses is likely to be reasonably small, surely much smaller than the many thousands of things that have reinforcing effects as a result of prior learning. If this number of values-as-criteria turns out to be small enough, the tasks of identifying them one-by-one and measuring them become considerably easier, and it also becomes easier to grapple with theoretical problems and problems of measurement concerning the organization of values into systems of values. And if the total number of human values is relatively small, and if all men everywhere possess them, comparative cross-cultural investigations of values would then become considerably easier also.

It seems, therefore, that there are compelling theoretical reasons for assuming that the study of a person's values is likely to be much more useful for social analysis than a study of the values that objects are said to have. Williams' argument makes sense on grounds of theoretical economy and social relevance and for other reasons as well. I have suggested elsewhere (Rokeach, 1968b), when comparing the relative power of the value concept as against other concepts, that by focusing upon a person's values "we would be dealing with a concept that is more central, more dynamic, more economical, a concept that would invite a more enthusiastic interdisciplinary collaboration, and that would broaden the range of the social psychologist's traditional concern to include problems of education and reeducation as well as problems of persuasion" (p. 159).

With the preceding considerations in mind, we may now offer the following definitions of what it means to say that a person has a *value* and a *value system*.

A *value* is an enduring belief that a specific mode of conduct or end-state of existence is personally or socially preferable to an opposite or converse mode of conduct or end-state of existence. A *value system* is an enduring organization of beliefs concerning preferable modes of conduct or end-states of existence along a continuum of relative importance.

These definitions, from which such terms as "ought," "should," and "conceptions of the desirable" have been deliberately excluded, are central to the present work. It should be noticed that the term "preferable" that is still present in the definition is not employed as a noun (i.e., a conception of the preferable) but as a predicate adjective, specifying that something is preferable to something else, that a particular mode of conduct or end-state of existence is preferable to an opposite mode or end-state. These definitions, and the elaborations of them that follow here, have affected for better or worse all the work to be reported in this book—on the measurement of values and value systems, their distribution in various segments of American society, their conceived antecedents and consequences, their relation to attitudes and behavior, the conditions under which values will undergo long-term change, and the cognitive and behavioral consequences that may be expected to follow from value change.

The two sections of this chapter which follow explain more fully what is intended by the above definitions of value and value system. In the course of doing so, related formulations that may be found in the literature will be reviewed and, moreover, compared with the present ones.

THE NATURE OF VALUES

A Value Is Enduring

If values were completely stable, individual and social change would be impossible. If values were completely unstable, continuity of human personality and society would be impossible. Any conception of human values, if it is to be fruitful, must be able to account for the enduring character of values as well as for their changing character.

It may be suggested that the enduring quality of values arises mainly from the fact that they are initially taught and learned in isolation from other values in an absolute, all-or-none manner. Such-and-such a mode of behavior or end-state, we are taught, is always desirable. We are not taught that it is desirable, for example, to be just a little bit honest or logical, or to strive for just a little bit of salvation or peace. Nor are we taught that such modes or end-states are sometimes desirable and sometimes not. It is the isolated and thus the absolute learning of values that more or less guarantees their endurance and stability.

Paradoxically, however, there is also a relative quality of values that must be made explicit if we are to come to grips with the problem of value change. As a child matures and becomes more complex, he is increasingly likely to encounter social situations in which several values rather than one value may come into competition with one another, requiring a weighing of one value against another—a decision as to which value is the more important. In this particular situation, is it better, for instance, to seek success or to remain honest, to act obediently or independently, to seek self-respect or social recognition? Gradually, through experience and a process of maturation, we all learn to integrate the isolated, absolute values we have been taught in this or that context into a hierarchically organized system, wherein each value is ordered in priority or importance relative to other values.

A simple analogy may be helpful here. Most parents tend to think they love each of their children in an absolute, unqualified manner. Yet, in a particular circumstance, a parent may nevertheless be forced to show a preference for one child over the others—for the one who is perhaps the most ill, or the most needful or frustrated, or the least able in school. Our values are like the children we love so dearly. When we think about, talk about, or try to teach one of our values to others, we typically do so without remembering the other values, thus regarding them as absolutes. But when one value is actually activated along with others in a given situation, the behavioral outcome will be a result of the relative importance of all the competing values that the situation has activated.

It is this relative conception of values that differentiates the present approach most distinctively from other approaches to the study of values.

A Value Is a Belief

Three types of beliefs have previously been distinguished (Rokeach, 1968b): descriptive or existential beliefs, those capable of being true or false; evaluative beliefs, wherein the object of belief is judged to be good or bad; and prescriptive or proscriptive beliefs, wherein some means or end of action is judged to be desirable or undesirable. A value is a belief of the third kind—a prescriptive or proscriptive belief. "A value is a belief upon which a man acts by preference" (Allport, 1961, p. 454).

Values, like all beliefs, have cognitive, affective, and behavioral components: (1) A value is a cognition about the desirable, equivalent to what Charles Morris (1956) has called a "conceived" value and to what Kluckhohn (1951) has called a "conception of the desirable." To say that a person has a value is to say that cognitively he knows the correct way to behave or the correct end-state to strive for. (2) A value is affective in the sense that he can feel emotional about it, be affectively for or against it, approve of those who exhibit positive instances and disapprove of those who exhibit negative instances of it. (3) A value has a behavioral component in the sense that it is an intervening variable that leads to action when activated.

A Value Refers to a Mode of Conduct or End-state of Existence

When we say that a person has a value, we may have in mind either his beliefs concerning desirable *modes of conduct* or desirable *end-states of existence*. We will refer to these two kinds of values as *instrumental* and *terminal* values. This distinction between means- and ends-values has been recognized by some philosophers (Lovejoy, 1950; Hilliard, 1950), anthropologists (Kluckhohn, 1951; Kluckhohn and Strodtbeck, 1961), and psychologists (English and English, 1958). But others have concentrated their attention more or less exclusively on one or the other kind of value. Thus, French and Kahn (1962), Kohlberg (1963), Piaget (1965), and Scott (1965) have for the most part concerned themselves with certain values representing idealized modes of conduct; Allport, Vernon and Lindzey (1960), Maslow (1959), Morris (1956), Rosenberg (1960), Smith (1969), and Woodruff (1942) have concerned themselves for the most part with certain values representing end-states.

This distinction between the two kinds of values—instrumental and terminal—is an important one that we cannot afford to ignore either in our theoretical thinking or in our attempts to measure values. For one thing, the total number of terminal values is not necessarily the same as the total number of instrumental values. For another, there is a functional relationship between instrumental and terminal values that cannot be ignored. These points will be developed more fully later on in this chapter.

Two Kinds of Terminal Values: Personal and Social. While there are no doubt many ways of classifying terminal values, one a priori classification that deserves to be singled out for special mention is that the terminal values may be self-centered or society-centered, intrapersonal or interpersonal in focus. Such end-states as salvation and peace of mind, for instance, are intrapersonal while world peace and brotherhood are interpersonal. It seems reasonable to anticipate that persons may vary reliably from one another in the priorities they place on such social and personal values; that their attitudes and behavior will differ from one another depending on whether their personal or their social values have priority; that an increase in one social value will lead to increases in other social values and decreases in personal values; and, conversely, that an increase in a personal value will lead to increases in other personal values and to decreases in social values.

Two Kinds of Instrumental Values: Moral Values and Competence Values. The concept of moral values is considerably narrower than the general concept of values. For one thing, moral values refer mainly to modes of behavior and do not necessarily include values that concern end-states of existence. For another, moral values refer only to certain kinds of instrumental values, to those that have an interpersonal focus which, when violated, arouse pangs of conscience or feelings of guilt for wrongdoing. Other instrumental values, those that may be called competence or self-actualization values, have a personal rather than interpersonal focus and do not seem to be especially concerned with morality. Their violation leads to feelings of shame about personal inadequacy rather than to feelings of guilt about wrongdoing. Thus, behaving honestly and responsibly leads one to feel that he is behaving morally, whereas behaving logically, intelligently or imaginatively leads one to feel that he is behaving competently. A person may experience conflict between two moral values (e.g., behaving honestly and lovingly), between two competence values (e.g., imaginatively and logically), or between a moral and a competence value (e.g., to act politely and to offer intellectual criticism).

One may raise the question whether there is a close connection between the two kinds of instrumental values, concerning morality and competence, and the two kinds of terminal values, concerning social and personal end-states. Will persons who place a higher priority on social end-states also place a higher priority on moral values? It might appear at first glance that the answer to this question is "yes," that the common thread running across the two kinds of terminal and instrumental values is an intrapersonal or interpersonal orientation. But further reflection suggests that such a simple one-to-one relationship cannot be expected. A person who is more oriented toward personal end-states may, for example, defensively place a higher priority on moral behavior. A person who is more oriented toward the social may also have a strong drive for personal competence (White, 1959), reflected in a greater priority placed on competence values. There is thus reason to doubt that there is any simple one-to-one connection between the two kinds of terminal and instrumental values.

On the "Oughtness" of Terminal and Instrumental Values. Virtually all writers have pointed to the "ought" character of values. "Ought," according to Heider (1958), is impersonal, relatively invariant, and interpersonally valid, deriving from what Kohler (1938) has called "objective requiredness." A person phenomenologically experiences "oughtness" to be objectively required by society in somewhat the same way that he perceives an incomplete circle as objectively requiring closure. The experience of "ought" can be "represented as a cognized wish or requirement of a suprapersonal objective order which has an invariant reality, and whose validity therefore transcends the point of view of any one person" (Heider, 1958, p. 222).

Can all terminal and instrumental values be claimed to have an "ought" character? It can be suggested that "oughtness" is more an attribute of instrumental than terminal values and more an attribute of instrumental values that concern morality than of those concerning competence. A person may feel more pressure upon him from the suprapersonal objective order to behave honestly and responsibly than to behave competently and logically; he may feel a greater pressure emanating from society to behave morally toward others than to seek such personal end-states as happiness and wisdom. It would thus appear that "oughtness" is not necessarily an attribute of all values. The more widely shared a value, the greater the societal demands placed upon us and therefore the greater the "oughtness" we experience. The "oughtness" of certain values is seen to originate within society, which demands that all of us behave in certain ways that benefit and do not harm others. It is an objective demand that we perceive society to place upon others no less than upon ourselves in order to ensure that all people live out their lives in a social milieu within which people can trust and depend upon one another. There seems to be little point in one person's behaving morally unless others also behave morally. In contrast, society seems somewhat less insistent in its demands concerning competent modes of behavior or concerning terminal end-states. Such values do not seem to be characterized by the same amount of "oughtness" that characterizes moral values. At most, there is a more subdued experience of "oughtness."

A Value Is a Preference as Well as a "Conception of the Preferable"

A good deal has been made of the distinction between the "desirable" and the "merely desired." Brewster Smith writes: "The more serious problem, which has yet to be solved in systematic research, is to distinguish dependably between values and preferences, between the desirable and the merely desired" (1969, p. 116). A value, as Kluckhohn defines it, is a "conception of the desirable," and not something "merely desired." This view of the nature of values suffers from the fact that it is extremely difficult to define "desirable." We are no better off and no further along talking about "conceptions of the desirable" than talking about values. More important, however, is that a conception of the "desirable which influences the selection from available modes, means, and ends of action" (Kluckhohn, 1951, p. 395) turns out, upon closer analysis, to represent a definable preference for something to something else. The something is a specific mode of behavior or end-state of existence; the something else is an opposite, converse, or contrary mode or end-state. Two mutually exclusive modes of behavior or end-states are compared with one another—for example, responsible and irresponsible behavior, or states of peace and war; one of the two is distinctly preferable to the other. Moreover, the person who prefers one believes that same one to be consensually preferred. A "conception of the desirable" thus seems to be nothing more than a special kind of preference—a preference for one mode of behavior over an opposite mode, or a preference for one end-state over an opposite end-state. Other kinds of preferences that do not implicate modes of behavior or end-states of existence, for instance, preferences for certain kinds of foods, would not qualify as "conceptions of the desirable."

There is also another sense in which a value represents a specific preference. A person prefers a particular mode or end-state not only when he compares it with its opposite but also when he compares it with other values within his value system. He prefers a particular mode or end-state to other modes or end-states that are lower down in his value hierarchy.

A Value Is a Conception of Something That Is Personally or Socially Preferable

If a person's values represent his "conceptions of the desirable" the question arises: desirable for whom? for himself? for others? When a person tells us about his values, it cannot be assumed that he necessarily intends them to apply equally to himself and to others. Consider, for example, the meaning of that familiar expression: "Children should be seen and not heard." Translated into the language of values, this statement apparently means to the person asserting it: "I believe it is desirable for children but not for adults to behave in certain ways." A person who informs us about his values may (or may not) intend to apply them differentially to young and old, men and women, blacks and whites, rich and poor, and so on.

Indeed, one of the most interesting properties that values seem to have is that they can be employed with such extraordinary versatility in everyday life. They may be shared or not shared and thus employed as single or double (or even triple) standards. They may be intended to apply equally to oneself and to others, to oneself but not to others, to others but not to oneself, to oneself more than to others, or to others more than to oneself. We know very little indeed about the conditions under which values might be so diversely employed. We may speculate, for example, that competitive conditions will encourage the employment of values as double standards, whereas cooperation will encourage their employment as single standards. A more systematic attack on this problem of single and double standards presents a major challenge to further theory and research on human values.

THE NATURE OF VALUE SYSTEMS

"It is the rare and limiting case," Robin Williams writes (1968, p. 287), "if and when a person's behavior is guided over a considerable period of time by one and only one value. . . . More often particular acts or sequences of acts are steered by multiple and changing clusters of values." After a value is learned it becomes integrated somehow into an organized system of values wherein each value is ordered in priority with respect to other values. Such a relative conception of values enables us to define change as a reordering of priorities and, at the same time, to see the total value system as relatively stable over time. It is stable enough to reflect the fact of sameness and continuity of a unique personality socialized within a given culture and society, yet unstable enough to permit rearrangements of value priorities as a result of changes in culture, society, and personal experience.

Variations in personal, societal, and cultural experience will not only generate individual differences in value systems but also individual differences in their stability. Both kinds of individual differences can reasonably be expected as a result of differences in such variables as intellectual development, degree of internalization of cultural and institutional values, identification with sex roles, political identification, and religious upbringing.

Number of Terminal and Instrumental Values

As already stated, the number of values human beings possess is assumed to be relatively small and, if small enough, it should be possible to identify and measure them. But how small? On various grounds—intuitive, theoretical, and empirical—we estimate that the total number of terminal values that a grown person possesses is about a dozen and a half and that the total number of instrumental values is several times this number, perhaps five or six dozen. On intuitive grounds, it seems evident that there are just so many end-states to strive for and just so many modes of behavior that are instrumental to their attainment. This number could not possibly run into the thousands or even into the hundreds, and, whatever the actual number may eventually turn out to be, it moreover seems evident that man possesses fewer terminal than instrumental values.

Certain theoretical considerations help bring us somewhat closer to an approximation of the total number of values. It can be argued that the total number of values is roughly equal to or limited by man's biological and social makeup and most particularly by his needs. How many needs do theorists say man possesses? Freud (1922) has proposed two, Maslow (1954) five, and Murray (1938) twenty-eight. These estimates suggest that the total number of terminal values may range somewhere between Freud's two and Murray's twenty-eight and that the total number of instrumental values may be several times this number.

Relation Between Instrumental and Terminal Values

What can be said about the relation between instrumental and terminal values? At this stage of theoretical thinking, it is safest to assume that they represent two separate yet functionally interconnected systems, wherein all the values concerning modes of behavior are instrumental to the attainment of all the values concerning end-states. This instrumentality is not necessarily a consciously perceived instrumentality, and there is not necessarily a one-to-one correspondence between any one instrumental value and any one terminal value. One mode of behavior may be instrumental to the attainment of several terminal values; several modes may be instrumental to the attainment of one terminal value.

Gorsuch (1970) has correctly pointed out that the "terminal-instrumental distinction may not go quite far enough" since "any value which is not the *ultimate* value could be considered an instrumental value." There is nevertheless a conceptual advantage to defining all terminal values as referring only to idealized end-states of existence and to defining all instrumental values as referring only to idealized modes of behavior. It may well be that one terminal value, so defined, is instrumental to another terminal value or that one instrumental value is instrumental to another instrumental value. Without ruling out such possibilities, it can nevertheless be suggested that the best strategy at an early stage of conceptualization is to conceive all instrumental values as modes of behavior that are instrumental to the attainment of all values concerning end-states of existence. In the final analysis, the fruitfulness of such a conceptualization will have to be tested empirically.

FUNCTIONS OF VALUES AND
VALUE SYSTEMS

One way to approach the question: what functions do values serve? is to think of values as standards that guide ongoing activities, and of value systems as general plans employed to resolve conflicts and to make decisions. Another way is to think of values as giving expression to human needs.

Values as Standards

Values are multifaceted standards that guide conduct in a variety of ways. They (1) lead us to take particular positions on social issues, and (2) predispose us to favor one particular political or religious ideology over another. They are standards employed (3) to guide presentations of the self to others (Goffman, 1959), and (4) to evaluate and judge, to heap praise and fix blame on ourselves and others. (5) Values are central to the study of comparison processes (Festinger, 1954; Latane, 1966); we employ them as standards to ascertain whether we are as moral and as competent as others. (6) They are, moreover, standards employed to persuade and influence others, to tell us which beliefs, attitudes, values, and actions of others are worth challenging, protesting, and arguing about, or worth trying to influence or to change.

Finally, (7) values are standards that tell us how to rationalize in the psychoanalytic sense, beliefs, attitudes, and actions that would otherwise be personally and socially unacceptable so that we will end up with personal feelings of morality and competence, both indispensable ingredients for the maintenance and enhancement of self-esteem. An unkind remark made to a friend, for example, may be rationalized as an honest communication; an inhibited sex life may be rationalized as a life guided by self-control; an act of aggression by a nation may be justified on the basis of one human value or another such as national security or the preservation of liberty. The process of rationalization, so crucial a component in virtually all the defense mechanisms, would be impossible if man did not possess values to rationalize with.

Indeed, the employment of values as standards is a distinctively human invention that is not shared with other species and is therefore one way of defining the difference between being human and nonhuman. It is an Aesopian language of self-justification on the one hand and of self-deception on the other (Frenkel-Brunswik, 1939) that enables us to maintain and enhance our self-esteem no matter how socially desirable or undesirable our motives, feelings, or actions may be. Values provide a basis for rational self-justification insofar as possible but also a basis for rationalized self-justification insofar as necessary. Either way, values serve to maintain and enhance self-esteem.

The proposition that values are standards that can be employed in so many different ways raises many difficult questions requiring further analysis and empirical investigation. Under what conditions will a value be employed as one kind of standard rather than another? Are there reliable individual differences in the way values are employed? Do some people typically employ certain values as standards of action, others as standards of judgment or evaluation, and yet others as standards to rationalize with?

Value Systems as General Plans for Conflict Resolution and Decision Making

Since a given situation will typically activate several values within a person's value system rather than just a single one, it is unlikely that he will be able to behave in a manner that is equally compatible with all of them. A given situation may, for example, activate a conflict between behaving independently and obediently or between behaving politely and sincerely; another situation may activate a conflict between strivings for salvation and hedonic pleasure or between self-respect and respect from others. A value system is a learned organization of principles and rules to help one choose between alternatives, resolve conflicts, and make decisions.

This is not to suggest, however, that a person's total value system is ever fully activated in any given situation. It is a mental structure that is more comprehensive than that portion of it that a given situation may activate. It is a generalized plan that can perhaps best be likened to a map or architect's blueprint. Only that part of the map or blueprint that is immediately relevant is consulted, and the rest is ignored for the moment. Different subsets of the map or blueprint are activated in different social situations.

Motivational Functions

If the immediate functions of values and value systems are to guide human action in daily situations, their more long-range functions are to give expression to basic human needs. Values have a strong motivational component as well as cognitive, affective, and behavioral components. Instrumental values are motivating because the idealized modes of behavior they are concerned with are perceived to be instrumental to the attainment of desired end-goals. If we behave in all the ways prescribed by our instrumental values, we will be rewarded with all the end-states specified by our terminal values. Terminal values are motivating because they represent the supergoals beyond immediate, biologically urgent goals. Unlike the more immediate goals, these supergoals do not seem to be periodic in nature; neither do they seem to satiate—we seem to be forever doomed to strive for these ultimate goals without quite ever reaching them.

But there is another reason why values can be said to be motivating. They are in the final analysis the conceptual tools and weapons that we all employ in order to maintain and enhance self-esteem. They are in the service of what McDougall (1926) has called the master sentiment—the sentiment of self-regard. A more detailed description of the ways in which values serve this sentiment of self-regard may profitably begin with certain formulations previously presented by Smith, Bruner, and White (1956) and by Katz (1960)—formulations that were primarily concerned with the functions of attitudes rather than values. According to these writers, attitudes serve value-expressive, adjustment, ego-defense, and knowledge functions. Any given attitude need not, however, serve all these functions and it may serve various combinations of these functions.

The present formulation differs somewhat from the preceding formulation about the four functions of attitudes. The value-expressive function is conceived here to be superordinate to the adjustment, ego defense, and knowledge functions because the content of values must concern itself with the relative desirability or importance of adjustment, ego defense, and knowledge. Put another way, the latter three functions represent nothing more than expressions or manifestations of different values that all people possess and hold dear to varying degrees. Thus, all of a person's attitudes can be conceived as being value-expressive, and all of a person's values are conceived to maintain and enhance the master sentiment of self-regard—by helping a person adjust to his society, defend his ego against threat, and test reality.

The Adjustive Function of Values. The *content* of certain values directly concerns modes of behavior and end-states that are adjustment- or utilitarian-oriented. For example, certain instrumental values concern the desirability of obedience, getting along well with others, politeness and self-control; certain terminal values concern the desirability of material comfort, success, prestige, and "law and order." Other values that may point somewhat less explicitly to the adjustment function are those stressing the importance of responsible and achievement-oriented behavior, and those emphasizing such terminal end-states as peace of mind and the security of self, family, and nation. Although we can all be assumed to possess such adjustment-oriented values, we differ in the importance we place on them relative to other values.

McLaughlin has suggested that adjustment-oriented values are really "pseudo-values" because they are "espoused by an individual as a way of adapting to group pressures" (1965, pp. 273-274). But the desirability of compliance (Kelman, 1961) to group pressures may be a genuine value in its own right, no less internalized than other values. We cannot expect, of course, that a person who values compliance will come right out and admit to possessing a value so baldly put, either to himself or to others. He would first have to transform it cognitively into values that are more socially and personally defensible—such values as success and getting along well with others.

The Ego-defensive Function of Values. Psychoanalytic theory suggests that values no less than attitudes may serve ego-defensive needs. Needs, feelings, and actions that are personally and socially unacceptable may be readily recast by processes of rationalization and reaction formation into more acceptable terms; values represent ready-made concepts provided by our culture to ensure that such justifications can proceed smoothly and effortlessly. All instrumental and terminal values may be employed to serve ego-defensive functions, but we can nevertheless single out certain values which especially lend themselves to such purposes. Research on the authoritarian personality (Adorno, *et al*, 1950) suggests that an overemphasis on such modes of behavior as cleanliness and politeness and on such end-states as family and national security may be especially helpful to ego defense. Research by many investigators (Allen and Spilka, 1967; Allport, 1954; Allport and Ross, 1967; Glock and Stark, 1965, 1966; Kirkpatrick, 1949; Lenski, 1961; Rokeach, 1969a, 19-69b) also suggests that religious values more often than not serve ego-defensive functions.

The Knowledge or Self-actualization Function of Values. Katz defines the knowledge function as involving "the search for meaning, the need to understand, the trend toward better organization of perception and belief to provide clarity and consistency" (1960, p. 170). Certain instrumental and terminal values explicitly or implicitly implicate this knowledge and, somewhat more broadly, the self-actualization function. Thus, people value such end-states as wisdom and a sense of accomplishment and such modes of behavior as behaving independently, consistently, and competently. We all possess such values, as we do adjustment and ego-defensive values, but again we differ in the priority we place on them. One person may, for example, attach greater importance to adjustment-oriented than knowledge-oriented values whereas another may reverse these priorities.

Higher- and Lower-order Values

Thus far in this discussion we have deliberately avoided labeling certain values as better or of a higher order than others. We have done so in the hope of demonstrating that it is possible to describe the values that people hold in a value-free manner. But it is now perhaps appropriate to suggest that values serving adjustive, ego-defensive, knowledge, and self-actualization functions may well be ordered along a continuum ranging from lower- to higher-order, as is suggested by Maslow's well-known hierarchical theory of motivation (1954). Different subsets of values may differentially serve Maslow's safety, security, love, self-esteem, and self-actualization needs. Maslow also (1959) speaks of B(being)-values and D(deficiency)-values, and in doing so he is again proposing that certain values are better, higher, more desirable for psychological fulfillment than others:

> For one thing, it looks as if *there were* a single ultimate value for mankind, a far goal toward which all men strive. This is called variously by different authors self-actualization, self-realization, integration, psychological health, individuation, autonomy, creativity, productivity, but they all agree that this amounts to realizing the potentialities of the person, that is to say, becoming fully human, everything that the person *can* become (p. 123).

The value concept employed by Maslow differs, however, in certain respects from the one presented here. He employs it in a manner that is more or less synonymous with the concept of need, drawing no distinction between instrumental and terminal values and dealing more with what has been identified here as end-states than with modes of behavior. These differences notwithstanding, Maslow's conception of higher- and lower-order values can be fruitfully employed. To the extent that a person's value system reflects a differential preoccupation with values that are adjustive, ego-defensive, and self-actualizing, we may say that he is operating at lower or higher levels. It remains to be seen to what extent various segments of American society can be thus described.

VALUES DISTINGUISHED FROM OTHER CONCEPTS

Brewster Smith (1969) is one of a small band of social scientists who has forcefully drawn our attention to the conceptual disarray of the value concept in the social sciences:

> But the increased currency of explicit value concepts among psychologists and social scientists has unfortunately not been accompanied by corresponding gains in conceptual clarity or consensus. We talk about altogether too many probably different things under one rubric when we stretch the same terminology to include the utilities of mathematical decision theory. . . , fundamental assumptions about the nature of the world and man's place in it. . . , ultimate preferences among life styles. . . , and core attitudes or sentiments that set priorities among one's preferences and thus give structure to a life. . . . And, at the same time, we are embarrassed with a proliferation of concepts akin to values: attitudes and sentiments, but also interests, preferences, motives, cathexes, valences. The handful of major attempts to study values empirically have started from different preconceptions and have altogether failed to link together to yield a domain of cumulative knowledge (pp. 97-98).

To reduce this conceptual disarray, we will attempt to distinguish the value concept employed here from various other concepts—attitude, social norm, need, trait, interest, and value orientation.

Values and Attitudes

Over the past fifty years empirically oriented social psychologists have paid considerably more attention to the theory and measurement of attitudes than to the theory and measurement of values. Despite the fact that many hundreds of value studies have accumulated over these years (Albert and Kluckhohn, 1958; Duffy, 1940; Dukes, 1955; Pittel and Mendelsohn, 1966) the ratio of attitude studies to value studies cited in *Psychological Abstracts* between 1961 and 1965 was roughly five or six attitude studies for every value study. There is little reason to think that this ratio would be markedly different for other years. This generally greater emphasis on attitudes has not arisen from any deep conviction that man's attitudes are more important determinants of social behavior than his values. Rather, it seems to have been forced upon us or to have evolved out of the more rapid development of methods for measuring attitudes, combined perhaps with a lack of clarity about the conceptual differences between values and attitudes and about their functional interconnections. Newcomb, Turner, and Converse, for instance, see values as but "special cases of the attitude concept" (1965, p. 45) and use the value concept only "informally"; Campbell regards the value and attitude concepts to be fundamentally similar (1963).

An attitude differs from a value in that an attitude refers to an organization of several beliefs around a specific object or situation (Rokeach, 1968a, 1968b). A value, on the other hand, refers to a single belief of a very specific kind. It concerns a desirable mode of behavior or end-state that has a transcendental quality to it, guiding actions, attitudes, judgments, and comparisons across specific objects and situations and beyond immediate goals to more ultimate goals. So defined, values and attitudes differ in a number of important respects. First, whereas a value is a single belief, an attitude refers to an organization of several beliefs that are all focused on a given object or situation. A Likert scale, for example, consists of a representative sample of beliefs all of which concern the same object or situation. When summed, it provides a single index of a person's favorable or unfavorable attitude toward an object or situation. Second, a value transcends objects and situations whereas an attitude is focused on some specified object or situation. Third, a value is a standard but an attitude is not a standard. Favorable or unfavorable evaluations of numerous attitude objects and situations may be based upon a relatively small number of values serving as standards. Fourth, a person has as many values as he has learned beliefs concerning desirable modes of conduct and end-states of existence, and as many attitudes as direct or indirect encounters he has had with specific objects and situations. It is thus estimated that values number only in the dozens, whereas attitudes number in the thousands.

Fifth, values occupy a more central position than attitudes within one's personality makeup and cognitive system, and they are therefore determinants of attitudes as well as of behavior. This greater centrality of values has occasionally been noted by others: "attitudes themselves depend on pre-existing social values" (Allport, 1961, pp. 802-803); "attitudes express values" (Watson, 1966, p. 215); "attitudes are functions of values" (Woodruff, 1942, p. 33). Closely related is the notion of perceived instrumentality (Peak, 1955; Carlson, 1956; Rosenberg, 1960; Homant, 1970; Hollen, 1972). A particular attitude object is perceived to be instrumental to the attainment of one or more values; a change in an attitude object's perceived instrumentality for one or more values should lead to a change in attitude; linking a particular attitude to more important values should make it more resistant to change than linking it to less important values (Ostrom and Brock, 1969; Edwards and Ostrom, 1969). Sixth, value is a more dynamic concept than attitude, having a more immediate link to motivation. "The now massive literature on attitudes serves to demonstrate nothing more clearly than that attitudes are not basic directive factors in behavior but that they are secondary to more personal characteristics . . ." (Woodruff, 1942, p. 33). If an attitude also has a motivational component, this is so only because the valenced (valued) attitude object or situation is perceived to be positively or negatively instrumental to value attainment. And, seventh, the substantive content of a value may directly concern adjustive, ego defense, knowledge or self-actualizing functions while the content of an attitude is related to such functions only inferentially.

Values and Social Norms

There are three ways in which values differ from social norms. First, a value may refer to a mode of behavior or end-state of existence whereas a social norm refers only to a mode of behavior. Second, a value transcends specific situations; in contrast, a social norm is a prescription or proscription to behave in a specific way in a specific situation. Navaho Indians, for example, should refrain from having ceremonials at the time of an eclipse of the moon (Kluckhohn, 1951, p. 413); Americans should stand respectfully at attention when the "Star Spangled Banner" is played at a public gathering but not when it is played in one's home. Third, a value is more personal and internal, whereas a norm is consensual and external to the person. Williams has aptly captured this difference between values and norms in the following passage:

> Values are standards of desirability that are more nearly independent of specific situations. The same value may be a point of reference for a great many specific norms; a particular norm may represent the simultaneous application of several separate values. . . . Values, as standards (criteria) for establishing what should be regarded as desirable, provide the grounds for accepting or rejecting particular norms (1968, p. 284).

Values and Needs

If some writers regard values and attitudes as more or less equivalent, others regard values and needs as equivalent. Maslow, for instance, refers to self-actualization both as a need and as a higher-order value (1959, 1964). Murray's list of needs (1938) is transformed into White's list of values (1951). French and Kahn (1962) point out that in some respects the properties of a value and of a need are similar. A person may want to do something but also feel that he ought to do it, since a value is not only a belief about what he ought to do but also a desire to do it.

> In other cases, perhaps especially in approach motivation, a single motivation system seems to have the properties of a need and also the properties of a value. In many people, for example, the achievement motive represents both something a person wants to do and something he feels he ought to do. For such social motives we assume that they have developed through a process of reinforcement similar to the development of moral values, and therefore, will have some of the same conceptual properties. Thus we shall try to define the conceptual properties of motives, including both needs and values, in such a way that they are not separated into a sharp dichotomy but instead have many conceptual properties in common (p. 11).

If values are indeed equivalent to needs, as Maslow and many others have suggested, then the lowly rat, to the extent that it can be said to possess needs, should to the same extent also be said to possess values. If such a view were adopted, it would be difficult to account for the fact that values are so much at the center of attention among those concerned with the understanding of human behavior and so little at the center of attention among those concerned with the understanding of animal behavior. That values are regarded to be so much more central in the one case than in the other suggests that values cannot altogether be identical to needs and perhaps that values possess some attributes that needs do not.

Man is the only animal that can be meaningfully described as having values. Indeed, it is the presence of values and systems of values that is a major characteristic distinguishing humans from infrahumans. Values are the cognitive representations and transformations of needs, and man is the only animal capable of such representations and transformations.

This proposition is not the whole story, however: Values are the cognitive representation not only of individual needs but also of societal and institutional demands. They are the joint results of sociological as well as psychological forces acting upon the individual—sociological because society and its institutions socialize the individual for the common good to internalize shared conceptions of the desirable; psychological because individual motivations require cognitive expression, justification, and indeed exhortation in socially desirable terms. The cognitive representation of needs as values serves societal demands no less than individual needs. Once such demands and needs become cognitively transformed into values, they are capable of being defended, justified, advocated, and exhorted as personally and socially desirable. For example, the need for sex which is so often repressed in modern society may be cognitively transformed as a value for love, spiritual union, or intimacy; needs for dependence, conformity, or abasement may be cognitively transformed into values concerning obedience, loyalty, or respect for elders; aggressive needs may be transformed into values concerning ambition, honor, family or national security. Needs may or may not be denied, depending on whether they can stand conscious personal and social scrutiny, but values need never be denied. Thus, when a person tells us about his values he is surely also telling us about his needs. But we must be cautious in how we infer needs from values because values are not isomorphic with needs. Needs are cognitively transformed into values so that a person can end up smelling himself, and being smelled by others, like a rose. Because infrahumans are incapable of such cognitive representations and transformations of needs, they cannot have values; consequently, a study of the laws governing animal behavior can lead to the discovery neither that men possess values nor that men use values as tools or weapons to preserve and enhance self-esteem.

Values and Traits

The concept of trait has had a long tradition in theory and research in personality (Allport, 1961; Rotter, 1954). It carries with it a connotation of human characteristics that are highly fixed and not amenable to modification by experimental or situational variation. It is difficult to locate experimental studies that are concerned with changes in traits, probably because the trait concept does not readily lend itself to alteration by experimental manipulation. About the only operations that one can easily perform on traits are to correlate them with other traits and to factor-analyze them. Nor are they amenable to change as a result of education or psychotherapy. Although psychotherapists often talk about the effects of therapy on habits, needs, attitudes, values, personality, and behavior, they rarely talk about the effects of therapy on traits.

A person's character, which is seen from a personality psychologist's standpoint as a cluster of fixed traits, can be reformulated from an internal, phenomenological standpoint as a system of values. Thus, a person identified from the "outside" as an authoritarian on the basis of his F-scale score can also be identified from the "inside" as one who places relatively high values on being obedient, clean, and polite and relatively low values on being broadminded, intellectual and imaginative. A person identified as an introvert on the basis of a score of an introversion-extroversion scale might identify himself as a person who cares more for wisdom and a life of the intellect and less for friendship, prestige, and being cheerful. A person we might identify as aggressive might see himself as merely ambitious, as one who cares about being a good provider for his family, and as one who cares about accomplishing something important in life.

A major advantage gained in thinking about a person as a system of values rather than as a cluster of traits is that it becomes possible to conceive of his undergoing change as a result of changes in social conditions. In contrast, the trait concept has built into it a characterological bias that forecloses such possibilities for change in advance. This very fixedness that has been built into the trait concept probably accounts for the fact that it has received so little attention from social psychologists and sociologists on the one hand and from students of behavior modification on the other.

Values and Interests

To Ralph Barton Perry (1954), a value is any object of interest, and the two are therefore identical concepts. Some writers have criticized the classical Study of Values (Allport, Vernon, and Lindzey, 1960) on the ground that it is primarily or solely a test of occupational interest (Duffy, 1940; McLaughlin, 1965).

An interest is but one of the many manifestations of a value, and therefore it has some of the attributes that a value has. An interest may be the cognitive representation of needs; it may guide action, evaluations of self and others, and comparisons of self with others. It may serve adjustive, ego-defense, knowledge, and self-actualization functions. But interest is obviously a narrower concept than value. It cannot be classified as an idealized mode of behavior or end-state of existence. It would be difficult to argue that an interest is a standard or that it has an "ought" character. It would, moreover, be difficult to defend the proposition that the interests that men have are relatively small in number and universally held or that they are organized into interest systems that serve as generalized plans for conflict resolution or decision making. Interests seem to resemble attitudes more than values, representing a favorable or unfavorable attitude toward certain objects (e.g., art, people, money) or activities (e.g., occupations).

Value Systems and Value Orientations

These two concepts seem at first glance to be more or less synonymous. But as employed by Kluckhohn and Strodtbeck, value orientation does not appear to be altogether interchangeable with the present notion of value system. Clyde Kluckhohn has defined a value orientation as "a set of linked propositions embracing both value and existential elements" (1951, p. 409). Florence Kluckhohn and Fred Strodtbeck (1961) measure value orientations operationally by asking respondents to rank-order alternative responses to each of five separate dimensions. Their value orientation refers to a pattern of rank-ordered results obtained within each of five separate dimensions and is not a rank ordering of the five dimensions with respect to one another. The notion of value system, in contrast, implies a rank ordering of terminal or instrumental values along a single continuum.

There is another reason this notion of value orientation is not considered to be equivalent to the notion of value system. The five separate dimensions formulated by Kluckhohn and by Kluckhohn and Strodtbeck—human nature is good or evil; subjugation to, harmony with, or mastery over nature; past, present, or future time perspective; being, being-in-becoming, or doing; linearity, collaterality, or individualism—seem to be somewhat far removed from what we ordinarily mean by a "conception of the desirable." A person may indeed believe that man is subjugated to nature but this circumstance does not necessarily imply that he has a value for "subjugation to nature," that he believes such a state of affairs to be desirable, or that man "ought" to be subjugated to nature. It appears that Kluckhohn and Strodtbeck's five dimensions can be more aptly described as basic philosophical orientations than as value orientations.

ANTECEDENTS AND CONSEQUENCES OF VALUES AND VALUE SYSTEMS

A major conceptual advantage of an approach wherein values are central is that we can with equal facility think of values as dependent or as independent variables. On the dependent side, they are a result of all the cultural, institutional, and personal forces that act upon a person throughout his lifetime. On the independent side, they have far-reaching effects on virtually all areas of human endeavor that scientists across all the social sciences may be interested in.

Antecedents of Values

Even if it were true that man possesses but a relatively small number of values, the number of theoretically possible variations in value systems is truly enormous, far more than needed to account for the rich differences that may exist among cultures, societies, institutional arrangements, and even among all individual personalities existing on planet Earth. A mere dozen and a half terminal values, for instance, can be arranged in order of importance in

18 factorial ways, which comes to over 640 trillion different ways. But it is extremely unlikely that all such theoretically possible permutations will actually be observed. We may expect that similarities in culture will sharply reduce the total number of possible variations to a much smaller number, shaping the value systems of large numbers of people in more or less similar ways. Further reductions in possible variations can moreover be expected within a given culture as a result of socialization by similar social institutions; similarities of sex, age, class, and race (Rokeach and Parker, 1970; Kohn, 1969); religious upbringing (Rokeach, 1969a, 1969b); political identification (Rokeach, 1968-1969); and the like.

We may also expect that similarities in personal experience and in the expression of individual needs will further reduce the total number of possible variations by shaping the value systems of many people in similar ways. Kluckhohn and Strodtbeck (1961) express a similar view concerning the limited number of variations in value orientations that are likely to be found within cultures:

> First, it is assumed that *there is a limited number of common human problems for which all peoples at all times must find some solution.* This is the universal aspect of value orientations because the common human problems to be treated arise inevitably out of the human situation. The second assumption is that *while there is variability in solutions of all the problems, it is neither limitless nor random but is definitely variable within a range of possible solutions.* The third assumption, the one which provides the main key to the later analysis of variation in value orientations, is that *all alternatives of all solutions are present in all societies at all times but are differentially preferred.* Every society has, in addition to its dominant profile of value orientations, numerous *variant* or *substitute profiles.* Moreover, it is postulated that in both the dominant and the variant profiles there is almost always a *rank ordering* of the preferences of the value-orientation alternatives (p. 10).

Consequences of Values

The values that are internalized as a result of cultural, societal, and personal experience are psychological structures that, in turn, have consequences of their own. These consequences have already been discussed in an earlier section on values as standards and therefore need be only briefly reiterated here. Values are determinants of virtually all kinds of behavior that could be called social behavior—of social action, attitudes and ideology, evaluations, moral judgments and justifications of self and others, comparisons of self with others, presentations of self to others, and attempts to influence others. Boiling all these down to a more succinct theoretical statement, it can perhaps be stated that values are guides and determinants of social attitudes and ideologies on the one hand and of social behavior on the other.

TOWARD A CLASSIFICATION OF HUMAN VALUES

"Much of the confusion in discussion about values," Kluckhohn writes, undoubtedly arises from the fact that one speaker has the general category in mind, another a particular limited type of value, still another a different specific type. We have not discovered any comprehensive classification of values. Golightly has distinguished essential and operational values; C. I. Lewis intrinsic, extrinsic, inherent, and instrumental values. The Cornell group speaks of asserted and operating values. Perry has discriminated values according to modalities of interest: Positive-negative, progressive-recurrent, potential-actual, and so on. There are various content classifications such as: hedonic, aesthetic, religious, economic, ethical, and logical. The best known of the content groupings is Spranger's (used in the Allport-Vernon test of values): theoretical, economic, aesthetic, social, political, and religious. The object to these content classifications is that they are culturebound. Ralph White has distinguished one hundred "general values" and twenty-five "political values" all with special references to Western culture (1951, p. 412).

Rather than burden the reader with yet another classification of values, I prefer to ask instead whether there might be some compelling theoretical basis for suggesting a systematic classification of values. A reasonable point of departure for such an attempt at classification is the observation that it is just as meaningful to speak of institutional values as of individual values. Every human value, as English and English have noted (1958), is a "social product" that has been transmitted and preserved in successive generations through one or more of society's institutions. We may define an institution as a social organization that has evolved in society and has been "assigned" the task of specializing in the maintenance and enhancement of selected subsets of values and in their transmission from generation to generation. Thus, religious institutions are institutions that specialize in furthering a certain subset of values that we call religious values; the family is an institution that specializes in furthering another subset of values; educational, political, economic, and legal institutions specialize in yet other subsets. The values that one institution specializes in are not necessarily completely different from those in which other institutions specialize. They may overlap and share certain values in common and thus reinforce each other's values, as in the case of the family and religious institutions. To the extent they do not overlap, however, they will compete with one another, as in the case of religious and secular institutions within a society that insists on separation of church and state.

If it is indeed the case that the maintenance, enhancement, and transmission of values within a culture typically become institutionalized, then an identification of the major institutions of a society should provide us with a reasonable point of departure for a comprehensive compilation and classification of human values.

SUMMARY

The following more extended definitions of a *value* and a *value system* are offered. To say that a person has a value is to say that he has an enduring prescriptive or proscriptive belief that a specific mode of behavior or end-state of existence is preferred to an oppositive mode of behavior or end-state. This belief transcends attitudes toward objects and toward situations; it is a standard that guides and determines action, attitudes toward objects and situations, ideology, presentations of self to others, evaluations, judgments, justifications, comparisons of self with others, and attempts to influence others. Values serve adjustive, ego-defensive, knowledge, and self-actualizing functions. Instrumental and terminal values are related yet are separately organized into relatively enduring hierarchical organizations along a continuum of importance.

REFERENCES

Adorno, T. W., Frenkel-Brunswik, E., Levinson, D. J., & Sanford, R. N. *The authoritarian personality*. New York: Harper, 1950.

Albert, E. M., & Kluckhohn, C. *A selected bibliography on values, ethics, and esthetics in the behavioral sciences and philosophy*. Glencoe, Ill.: Free Press, 1959.

Allen, R. O., & Spilka, B. Committed and consensual religion: A specification of religion-prejudice relationships. *Journal for the Scientific Study of Religion, 1967, 6*, 191-206.

Allport, G. W. *Pattern and growth in personality*. New York: Holt, Rinehart, & Winston, 1961.

Allport, G. W., & Ross, J. M. Personal religious orientation and prejudice. *Journal of Personality and Social Psychology*, 1967, 5, 432-443.

Allport, G. W., Vernon, P. E., & Lindzey, G. *A study of values*. Boston: Houghton Mifflin, 1960.

Campbell, D. T. Social attitudes and other acquired behavioral dispositions. In S. Koch (Ed.), *Psychology: A study of a Science*, Vol. 6. New York: McGraw-Hill, 1963.

Campbell, D. T., & Stanley, J. C. *Experimental and quasi-experimental designs for research*. Chicago: Rand McNally, 1963.

Carlson, E. R. Attitude change through modification of attitude structure. *Journal of Abnormal and Social Psychology*, 1956, 52, 256-261.

Duffy, E. A. A critical review of investigations employing the Allport-Vernon study of values and other tests of valuative attitudes. *Psychological Bulletin*, 1940, 37, 597-612.

Dukes, W. F. Psychological studies of values. *Psychological Bulletin*, 1955, 52, 24-50.

Edwards, J. D., & Ostrom, T. M. Value-bonded attitudes: Changes in attitude structure as a function of value bonding and type of communication discrepancy. *Proceedings, 77th Annual Convention, American Psychological Association*, 1969, 413-414.

English, H. B., & English, A. C. *A comprehensive dictionary of psychological and psychoanalytic terms*. New York: Longmans, Green, 1958.

French, J. R. P., & Kahn, R. L. A programmatic approach to studying the industrial environment and mental health. *Journal of Social Issues*, 1962, 18, 1-47.

Frenkel-Brunswik, E. Mechanisms of self-deception. *Journal of Social Psychology* (S.P.S.S.I. Bulletin), 1939, 10, 409-420.

Freud, S. *Beyond the pleasure principle*. London: Hogarth Press, 1922.

Glock, C. Y., & Stark, R. *Religion and society in tension*. Chicago: Rand McNally, 1965.

Goffman, E. *The presentation of self in everyday life*. Garden City, N.Y.: Doubleday, 1959.

Gorsuch, R. L. Rokeach's approach to value systems and social compassion. *Review of Religious Research*, 1970, 11, 139-143.

Handy, R. *The measurement of values*. St. Louis, Mo.: Green, 1970.

Heider, F. *The psychology of interpersonal relations*. New York: Wiley, 1958.

Hilliard, A. L. *The forms of value: The extension of a hedonistic axiology*. New York: Columbia University Press, 1950.

Hollen, C. C. Value change, perceived instrumentality, and attitude change. Unpublished Ph.D. dissertation, Michigan State University Library, 1972.

Homant, R. Values, attitudes, and perceived instrumentality. Unpublished Ph.D. dissertation, Michigan State University Library, 1970.

Jones, E. E., & Gerard, H. B. *Foundations of social psychology*. New York: Wiley, 1967.

Katz, D. The functional approach to the study of attitudes. *Public Opinion Quarterly*, 1960, **24**, 163-204.

Katz, D., & Stotland, E. A preliminary statement to a theory of attitude structure and change. In S. Koch (Ed.), *Psychology: A study of a science*. New York: McGraw-Hill, 1959.

Kelman, H. C. Processes of opinion change. *Public Opinion Quarterly*, 1961, **25**, 57-78.

Kirkpatrick, C. Religion and humanitarianism: A study of institutional implications. *Psychological Monographs*, 1949, **63** (Whole No. 304).

Kluckhohn, C. Values and value orientations in the theory of action. In T. Parsons & E. A. Shils (Eds.), *Toward a general theory of action*. Cambridge: Harvard University Press, 1951.

Kluckhohn, F. & Strodtbeck, F. L. *Variations in value orientation*. Evanston, Ill.: Row, Peterson, 1961.

Kohlberg, L. The development of children's orientations toward a moral order. I Sequences in the development of mental thought. *Vita Humana*, 1963, **6**, 11-33.

Kohler, W. *The place of value in a world of facts*. New York: Liveright, 1938.

Kohn, M. L. *Class and conformity: A study in values*. Homewood, Ill.: Dorsey, 1969.

Latane, B. (Ed.) Studies in social comparison. *Journal of Experimental and Social Psychology*, 1966, Supplement 1 (Whole issue).

Lenski, G. *The religious factor*. Garden City, N.Y.: Doubleday, 1961.

Lewis, C. I. *An analysis of knowledge and valuation*. LaSalle, Ill.: Open Court, 1962.

Lovejoy, A. O. Terminal and adjectival values. *Journal of Philosophy*, 1950, **47**, 593-608.

Maslow, A. H. (Ed.) *New knowledge in human values*, New York: Harper & Row, 1959.

Maslow, A. H. *Religions, values, and peak experiences*. Columbus, Ohio: Ohio State University Press, 1964.

McDougall, W. *An introduction to social psychology*. Boston: John W. Luce, 1926.

McLaughlin, B. Values in behavioral science. *Journal of Religion and Health*, 1965, **4**, 258-279.

Morris, C. W. *Varieties of human value*. Chicago: University of Chicago Press, 1956.

Murray, H. A. *Explorations in personality: A clinical and experimental study of fifty men of college age*. New York: Oxford University Press, 1938.

Newcomb, T. M., Turner, R. H., & Converse, P. E. *Social psychology*. New York: Holt, Rinehart, & Winston, 1965.

Ostrom, T. M., & Brock, T. C. Cognitive bonding to central values and resistance to a communication advocating change in policy orientation. *Journal of Experimental Research in Personality*, 1969, **4**, 42-50.

Peak, H. Attitude and motivation. In M. R. Jones (Ed.), *Nebraska symposium on motivation*. Lincoln, Neb.: University of Nebraska Press, 1955.

Perry, R. B. *Realms of value: A critique of human civilization*. Cambridge: Harvard University Press, 1954.

Piaget, J. *The moral judgment of the child*. New York: Free Press, 1965.

Pittel, S. M., & Mendelsohn, G. A. Measurement of moral values: A review and critique. *Psychological Bulletin*, 1966, **66**, 22-35.

Rokeach, M. A theory of organization and change within value-attitude systems. *Journal of Social Issues*, 1968, **24**, 13-33.(a)

Rokeach, M. *Beliefs, attitudes, and values.* San Francisco: Jossey-Bass, 1968.(b)

Rokeach, M. The role of values in public opinion research. *Public Opinion Quarterly,* 1968-69, **32,** 547-559.

Rokeach, M. Value systems in religion. *Review of Religious Research,* 1969, **11,** 3-23.(a)

Rokeach, M. Religious values and social compassion. *Review of Religious Research,* 1969, **11,** 24-38.(b)

Rokeach, M., & Parker, S. Values as social indicators of poverty and race relations in America. *The Annals of the American Academy of Political and Social Science,* 1970, **388,** 97-111.

Rosenberg, M. J. An analysis of affective-cognitive consistency. In M. J. Rosenberg *et al.* (Eds.), *Attitude organization and change.* New Haven: Yale University Press, 1960.

Rotter, J. *Social learning and clinical psychology.* New York: Prentice-Hall, 1954.

Scott, W. A. *Values and organizations.* Chicago: Rand McNally, 1965.

Skinner, B. F. *Beyond freedom and dignity.* New York: Knopf, 1971.

Smith, M. B. *Social psychology and human values.* Chicago: Aldine, 1969.

Smith, M. B., Bruner, J. S., & White, R. W. *Opinions and personality.* New York: Wiley, 1956.

Thomas, W. I., & Znaniecki, F. *The Polish peasant in Europe and America,* Vol. 1. Boston: Badger, 1918-20.

Watson, G. *Social psychology: Issues and insights.* Philadelphia: Lippincott, 1966.

White, R. K. *Value analysis: Nature and use of the method.* Ann Arbor, Mich.: Society for the Psychological Study of Social Issues, 1951.(a)

White, R. K. *Value-analysis, the nature and use of the method.* Glen Gardner, N.J.: Libertarian Press, 1951.(b)

White, R. W. Motivation reconsidered: The concept of competence. *Psychological Review,* 1959, **66,** 297-333.

Williams, R. M. Values. In E. Sills (Ed.), *International encyclopedia of the social sciences.* New York: Macmillan, 1968.

Woodruff, A. D. Personal values and the direction of behavior. *School Review,* 1942, **50,** 32-42.

Woodruff, A. D., & DiVesta, F. J. The relationship between values, concepts, and attitudes. *Educational and Psychological Measurement,* 1948, **8,** 645-659.

AGE DIFFERENCES IN VALUES

Milton Rokeach

The results to be reported here are for several samples of Americans ranging in age from 11 at one extreme to 70 and over at the other. The category "70 and over" consists of 169 respondents in the national sample of whom 37 were between 80 and 90, and 3 were over 90. The findings shown in Tables 1 through 4 are from three samples. The national NORC sample, already described, provided the data for adult Americans over twenty-one. The college sample was obtained from introductory psychology classes at Michigan State University; most of the students, about evenly divided in sex, ranged in age from eighteen to twenty-one. Needless to say, this sample can hardly be considered to be representative of American youth of the same age. It is essentially a middle-class sample of college students recruited mainly from the Midwest. [Additional data on junior college freshmen in California have recently been reported by Brawer (1971).] The eleven-, thirteen-, fifteen-, and seventeen-year-old groups were obtained by Drs. Robert Beech and Aileen Schoeppe (1970), from the public schools of New York City, and are presented here with their permission. Each age group consists of approximately equal numbers of boys and girls tested in middle-class and lower-class areas. The reader is therefore strongly cautioned to take the nature of these three samples into account when interpreting the findings shown in Tables 1 through 4. Because of the voluminousness of these findings, the terminal and instrumental value medians are shown separately in Tables 1 and 2, and the composite rank orders are shown in 3 and 4. It should, moreover, be noted that significance of differences is shown separately for the national sample (Median test) and for the four adolescent groups (one-way analysis of variance). No attempt was made to test for significance of differences across the total age range because the three samples do not come from the same population.

At least 30 of the 36 values show significant age differences, either in the adolescent or national sample or in both samples. The general impression gained from an inspection of the data is one of continuous value change from early adolescence through old age with the presence of several generation gaps rather than just one. To aid in the interpretation of these findings, the composite rankings were plotted against age for each of the 36 values separately, and an attempt was made to identify those values that showed similar patterns of development. By this process, the developmental patterns observed for each of

246

Table 1. Terminal Value Medians for Eleven Age Groups

Terminal Values	11	13	15	17	p	College	20's	30's	40's	50's	60's	70's	p
N =	190	183	186	193		298	267	298	280	236	159	169	
A comfortable life	8.5	9.2	9.2	10.2	—	12.1	9.5	10.1	8.7	9.1	8.0	7.7	.05
An exciting life	10.6	10.8	12.0	11.9	—	11.6	15.3	14.6	15.5	15.4	15.4	16.2	.01
A sense of accomplishment	13.2	11.5	11.8	9.4	.001	7.7	9.0	8.9	8.9	8.7	9.3	10.2	—
A world at peace	3.2	3.2	2.6	3.4	.001	8.4	3.8	4.2	3.1	3.0	2.9	2.8	.01
A world of beauty	8.2	11.5	12.2	12.7	.001	13.8	13.9	13.7	13.7	13.3	13.6	12.9	—
Equality	6.9	6.4	5.4	5.4	—	10.6	8.0	8.2	9.0	9.3	10.4	7.7	—
Family security	5.2	3.5	5.3	7.5	.001	8.4	4.0	3.2	3.3	3.7	4.5	5.6	.01
Freedom	4.6	5.1	4.1	5.4	.05	5.1	4.3	5.4	5.9	6.2	5.7	6.6	.01
Happiness	6.8	7.2	8.6	7.4	—	6.2	7.0	7.5	7.3	8.1	8.1	8.0	—
Inner harmony	11.3	13.7	12.4	9.3	.001	8.3	11.0	9.9	10.0	10.7	10.1	11.0	—
Mature love	9.4	10.9	10.1	10.0	—	7.3	10.0	10.8	12.9	13.1	14.4	15.5	.01
National security	12.6	12.2	9.6	14.2	.001	13.6	9.9	11.4	9.7	8.9	7.4	8.2	.01
Pleasure	11.4	10.9	13.2	12.6	.01	14.0	14.5	15.3	14.8	14.5	14.4	13.4	.01
Salvation	14.6	13.9	13.8	15.9	.01	13.1	9.7	10.0	8.6	6.8	9.0	6.4	.05
Self-respect	12.2	11.6	9.7	8.4	.001	6.7	8.4	7.0	7.9	8.0	7.6	7.6	—
Social recognition	13.8	12.7	12.9	12.2	.05	13.6	14.5	14.7	14.4	14.2	13.6	14.5	—
True friendship	6.5	7.0	8.4	8.5	.01	7.8	10.0	9.9	9.7	8.6	8.7	7.8	.01
Wisdom	9.9	9.1	8.0	7.3	.01	6.4	7.9	7.1	7.9	8.2	8.7	9.0	.05

Table 2. Instrumental Value Medians for Eleven Age Groups

Instrumental Values	N =	11	13	15	17	p	College	20's	30's	40's	50's	60's	70's	p
		190	183	186	193		298	267	298	280	236	159	169	
Ambitious		10.0	7.3	7.1	6.2	.001	7.0	7.6	6.4	5.9	6.4	6.1	6.3	—
Broadminded		11.7	13.2	11.6	8.4	.001	6.7	8.2	8.2	7.5	6.3	7.1	6.6	—
Capable		10.2	9.3	9.5	8.8	—	9.4	10.5	9.4	9.1	9.7	9.0	9.7	.05
Cheerful		6.5	9.8	11.2	11.1	.001	11.4	10.9	10.6	10.3	9.3	8.5	8.9	.01
Clean		8.1	8.9	9.6	9.9	—	14.2	8.3	10.1	8.7	8.8	7.5	8.1	.05
Courageous		9.3	9.3	8.8	7.9	—	8.4	8.7	7.4	7.5	7.6	8.6	7.6	—
Forgiving		7.6	8.6	7.9	9.5	.01	9.7	7.6	8.1	6.9	7.4	7.1	6.0	.01
Helpful		6.5	6.5	8.1	9.1	.001	10.6	9.3	8.8	8.0	8.1	8.1	7.2	.01
Honest		4.6	4.5	4.5	5.9	.05	4.2	3.4	2.7	3.4	3.4	3.4	3.5	—
Imaginative		12.5	14.3	14.3	12.9	.01	11.4	15.0	15.1	15.3	15.6	16.0	15.4	—
Independent		10.5	11.8	9.8	8.8	.05	8.1	10.4	9.7	10.8	11.2	10.5	9.9	—
Intellectual		12.5	12.1	11.8	10.9	—	9.2	12.3	13.1	12.8	13.1	13.6	13.6	—
Logical		13.1	15.2	14.0	12.7	.001	10.1	13.9	13.6	14.0	14.6	15.2	14.3	.01
Loving		4.8	5.7	5.6	7.2	.05	7.7	8.2	9.1	9.9	9.8	11.2	11.1	—
Obedient		11.3	10.1	11.9	12.7	.05	15.3	13.0	13.8	13.0	13.7	13.0	13.2	—
Polite		9.1	8.8	10.0	10.9	—	12.9	10.3	10.8	12.2	10.5	10.1	10.7	.01
Responsible		9.3	7.8	7.5	7.4	.01	6.0	6.0	6.3	6.1	6.9	7.7	8.1	.01
Self-controlled		10.5	9.7	10.2	10.7	—	8.8	9.6	8.8	10.0	9.9	9.9	10.0	—

Table 3. Composite Rank Orders of Terminal Values for Eleven Age Groups.

Terminal Values	N = 11	13	15	17	p	College	20's	30's	40's	50's	60's	70's	p
	190	183	186	193		298	267	298	280	236	159	169	
A comfortable life	8	8	8	12	—	13	9	12	8	11	6	6	.05
An exciting life	11	9	13	13	—	12	18	16	18	18	18	18	.01
A sense of accomplishment	16	13	12	10	.001	6	8	8	9	9	11	12	—
A world at peace	1	1	1	1	—	10	1	2	1	1	1	1	.01
A world of beauty	7	12	14	16	.001	17	15	15	15	15	15	14	—
Equality	6	4	4	2	—	11	6	7	10	12	13	7	—
Family security	3	2	3	6	.001	9	2	1	2	2	2	2	.01
Freedom	2	3	2	3	.05	1	3	3	3	3	3	4	.01
Happiness	5	6	7	5	—	2	4	6	4	6	7	9	—
Inner harmony	12	17	15	9	.001	8	14	9	13	13	12	13	.01
Mature love	9	11	11	11	—	5	13	13	14	14	16	17	.01
National security	15	15	9	17	.001	16	11	14	11	10	4	10	.01
Pleasure	13	10	17	15	.01	18	16	18	17	17	17	15	.01
Salvation	18	18	18	18	.01	14	10	11	7	4	10	3	.05
Self-respect	14	14	10	7	.001	4	7	4	5	5	5	5	—
Social recognition	17	16	16	14	.05	15	17	17	16	16	14	16	.01
True friendship	4	5	6	8	.01	7	12	10	12	8	9	8	.01
Wisdom	10	7	5	4	.01	3	5	5	6	7	8	11	.05

Table 4. Composite Rank Orders of Instrumental Values for Eleven Age Groups

Instrumental Values	11	13	15	17	p	College	20's	30's	40's	50's	60's	70's	p
N=	190	183	186	193		298	267	298	280	236	159	169	
Ambitious	10	4	3	2	.001	4	4	3	2	3	2	3	—
Broadminded	15	16	14	6	.001	3	6	6	5	2	4	4	—
Capable	11	9	8	7	—	10	13	10	9	10	10	10	.05
Cheerful	3	12	13	15	.001	15	14	13	12	9	8	9	.01
Clean	6	8	9	11	—	17	7	12	8	8	5	7	.05
Courageous	8	10	7	5	—	7	8	4	6	6	9	6	—
Forgiving	5	6	5	10	.01	11	3	5	4	5	3	2	.01
Helpful	4	3	6	9	.001	13	9	7	7	7	7	5	.01
Honest	1	1	1	1	.05	1	1	1	1	1	1	1	—
Imaginative	16	17	18	18	.01	14	18	18	18	18	18	18	—
Independent	12	14	10	8	.05	6	12	11	13	14	13	11	—
Intellectual	17	15	15	14	—	9	15	15	15	15	16	16	—
Logical	18	18	17	16	.001	12	17	16	17	17	17	17	.01
Loving	2	2	2	3	.05	5	5	9	11	11	14	14	—
Obedient	14	13	16	17	.05	18	16	17	16	16	15	15	.01
Polite	7	7	11	13	—	16	11	14	14	13	12	13	—
Responsible	9	5	4	4	.01	2	2	2	3	4	6	8	.01
Self-controlled	13	11	12	12	—	8	10	8	10	12	11	12	—

the 36 values were reduced to 14, and these 14 developmental patterns are described below.

Developmental Pattern 1: A Sense of Accomplishment, Wisdom, Responsible

All values of this subset seem to concern self-realization and show an inverted U-shaped pattern. All three are ranked relatively low or in the middle of the value scale in early adolescence, then increase gradually in importance through the adolescent and college years, and then gradually become less important in the decades beyond. They end up about as unimportant in old age as they were to begin with in early adolescence. The results for *wisdom* are especially noteworthy since they contradict the widely held belief that *wisdom* is especially valued in old age.

Developmental Pattern 2: Imaginative, Intellectual, Logical, Inner Harmony

This developmental pattern seems to be closely related to Pattern 1. All four of these personal values are ranked uniformly low to begin with by adolescent groups, then increase suddenly to become moderately important during the college years, and then drop quickly again to a position of relatively low priority for all age groups beyond. The results for *logical* are of special relevance to social-psychological theories concerned with cognitive consistency. *Logical* is ranked sixteenth to eighteenth down the list by adolescents, rises to twelfth among college students, and then drops once again toward the bottom among adult Americans (Table 4). College professors, who typically place a high value on being *logical* (ranking it anywhere from 5th to 8th in importance), seem in their formulation of consistency and balance theories to have projected their own values for being *logical* onto others, perhaps on the simple assumption that others value what academicians value. But this is obviously not so. The average adolescent and adult American care little indeed about being *logical*, and college students care only a bit more about it. Generalizations about human behavior based upon experiments with college students therefore seem to be unwarranted when the experimental task presupposes logical inconsistency among cognitive components as a major motivating force for attitude or behavioral change.

Developmental Pattern 3: A World of Beauty, True Friendship, Polite

Although it is difficult to discern the common element in these three values, they nonetheless show a similar developmental pattern. They all start out as relatively important in early adolescence but become increasingly less important in the college years and the twenties, at which point they more or less level off to remain relatively unimportant in the following decades. The results for *a world of beauty* deserve special mention in an era of increasing apprehension about quality of life, pollution, and ecology. It is ranked seventh in importance by eleven-year-olds, then declines steadily during the adolescent years to a position of seventeenth down the list by students in college, and there it remains toward the bottom of the terminal value hierarchy—for all age groups beyond. It would seem as if the socialization process had somehow destroyed the young adolescent's initial appreciation of beauty by replacing it with other values that are deemed more important. Perhaps our present concern with ecology will reverse *a world of beauty's* generally low position in the American value hierarchy in the years to come.

Developmental Pattern 4: Obedience

Somewhat similar to Pattern 3 is the one shown by *obedient*, except for the fact that it starts out lower down the list of instrumental values for eleven-year-olds—fourteenth. It then declines gradually to eighteenth for college students and then gradually rises decade after decade to a position of fifteenth in importance.

Developmental Pattern 5: An Exciting Life, Pleasure

These two hedonistic values are moderately important during early adolescence and then become less and less important. They are virtually at the bottom of the value hierarchy for college students or for people in their twenties, and they are ranked at or near the bottom of the value hierarchy in the decades that follow.

Developmental Pattern 6: Self-respect, Ambitious, Broadminded

These three values, which seem to concern self-realization, are all relatively unimportant, perhaps because they are still meaningless to young adolescents. All three increase steadily during the adolescent years to assume a position of major importance in late adolescence and the college years. They then level off and continue to remain important.

Developmental Pattern 7: Loving

Of all the 36 values *loving* shows the most clearcut linear relation to age. It starts out second in importance for eleven-year-olds and then declines more or less linearly to fourteenth for people beyond seventy. It would thus seem that love is important for young people but, surprisingly, becomes less and less important as people grow older.

Developmental Pattern 8: Mature Love

This value shows a developmental pattern somewhat similar to that found for *loving*, except for the fact that it starts out ninth in importance for young adolescents and declines steadily to seventeenth for people in their seventies and beyond. The only exception is its sudden rise to fifth place in the college years and its sudden decline thereafter. It would seem that *mature love* is an important value mainly for young middle-class college students.

Developmental Pattern 9: A World at Peace, Family Security, Capable

All three of these rather diverse values are more or less equally valued across all age groups except that they all show a sharp decrease in importance in late adolescence and during the college years and then an equally sharp increase. This pattern is most evident for *a world at peace* which is at the top of the terminal value hierarchy for all groups, except that it declines to tenth position for those in college. This devaluation is possibly due to the draft status of college students compared with other groups. Perhaps it is an attempt to rationalize one's continuing in college and avoiding the draft. Similar patterns are noted for *family security* and *capable*. The devaluing of *family security* by those in late adolescence probably reflects the fact that love and marriage are more focal at this age than raising a family. The developmental pattern for *capable*, however, is more surprising. From seventeen through the twenties the importance of being competent declines. From then on until the forties it becomes more important, and then it levels off.

Developmental Pattern 10: A Comfortable Life, Cheerful, Clean, Forgiving, Helpful

All these are rather conventional values and they all show a gradual decrease in importance during the adolescent years followed, in later decades, by a gradual, though somewhat uneven, increase in importance. Since all these values are significant indicators of socioeconomic status, their developmental pattern can be interpreted as reflecting a lesser concern with socioeconomic status in late adolescence and an increasing concern with it later on, after reaching the age of marriage. All these class-related values, with the exception of *helpful*, seem to become increasingly important in the later years of life, as issues of economic security and religion become more salient.

Developmental Pattern 11: Equality, Independent

Both of these values show a rather similar complex, undulating pattern of development. They both increase gradually in importance during adolescence, then show a sudden decrease during the college years or the twenties, then another increase, then a gradual decrease, and culminate in yet another increase in importance from the sixties to the seventies.

Developmental Pattern 12: Salvation

This value is evidently one of extremely low priority throughout adolescence, as evidenced by the fact that it is uniformly ranked at the bottom of the terminal value hierarchy during the adolescent years. It then increases in importance to fourteenth during the college years and continues to become more important in succeeding decades, ranking third among those beyond seventy. This smooth developmental trend is, however, interrupted in the sixties when *salvation* drops inexplicably to tenth in importance. It is difficult to say without further research whether this particular finding is due to chance. Whether or not this will turn out to be the case, the overall developmental trend is unmistakable: *Salvation* becomes increasingly important with advancing age.

Developmental Pattern 13: National Security

This value is a relatively unimportant one for eleven- and thirteen-year-olds, ranking fifteenth among the terminal values. It increases to ninth in importance for fifteen-year-olds and then plummets sharply to sixteenth or seventeenth for seventeen-year-olds and for those in college. It then increases steadily in importance with each decade from the twenties to the sixties at which time it is ranked fourth from the top. It then drops back to tenth in the years beyond the seventies. It seems especially ironic in this era of student protest that *national security* is near the bottom of the terminal value hierarchy for those who are most eligible for military service and edges close to the top of the value hierarchy for those in their later years, prior to retirement, when they are at the height of their identification with and participation in the Establishment. Differences in the perceived importance of *national security* were surely a major component of the generation gap in the late 1960's.

Developmental Pattern 14: Freedom, Happiness, Social Recognition, Courageous, Honest, Self-controlled

Finally, these six values show relatively little fluctuation with age. *Honest* seems to be the most stable of all the 36 values, its composite ranking being first for all age groups without exception. Next most stable as an American value is *freedom*, which is ranked among the top four values for all age groups. The remaining four values—*happiness, social recognition, courageous,* and *self-controlled*—show somewhat more fluctuation with age but the differences are not statistically significant, at least when the adolescent groups are compared or when the age groups in the national adult sample are compared.

Before leaving the issue of value development it should perhaps be stated that the 14 value patterns discussed here are purely descriptive in nature; they are based upon a visual inspection of data obtained from adolescents in New York City, college students in Michigan, and adult Americans from a national sample. Further research is obviously needed to determine whether the value patterns described here are replicable with more representative sampling procedures across the total age range, whether the number of value patterns can be reduced to a still smaller number, and further research is needed in order to account for the developmental patterns observed. Most particularly, further research must determine to what extent the generational differences observed are a function of maturation and to what extent they are a function of changes in generational norms concerning socialization. Even though it is not possible (without longitudinal data) to assess their relative contributions, it is safe to assume that both factors have influenced the observed results.

Perhaps most illuminating from a theoretical standpoint is the finding that value changes take place not only during adolescence but throughout life. Although the data do not permit us to generalize below age eleven, value development is, at least in large part, a continuing maturational process of change from birth to death and does not stop arbitrarily with the end of the period of psychosexual development that Freud speaks of. The results are therefore more in line with Erikson's view of development (1950) rather than with Freud's more classical view.

In the context of the present discussion of age-related values, it is perhaps appropriate to comment also on the relation between values, as conceived and measured here, and moral development, as conceived and measured by Kohlberg (1964). This relation has been investigated by McLellan (1970) with 78 male subjects from the seventh, ninth, and eleventh grades. These subjects were tested with Kohlberg's Moral Judgment instrument and the Rokeach Value Survey. McLellan found that two values especially, *freedom* and *obedient*, discriminated among Kohlberg's moral stages when age was held constant. As would be predicted from Kohlberg's theory, *freedom* rankings were found to be highest at stage 2 (the instrumentalist relationist orientation) and at stage 5 (the social contract legalistic orientation), and that *obedient* rankings were highest at stage 1 (the punishment and obedience orientation) and at stage 4 (the rigid rule orientation). McLellan's findings suggest that a more extensive study of the relation between stages of moral development and the organization of values would be rewarding.

RELIGIOUS VALUES

Detailed reports of value similarities and differences among the religious, the less religious, and the nonreligious have been reported elsewhere (Rokeach, 1969a, 1969b, 1970a, 1970b) and will therefore only be summarized here. All religious groups are similar in considering *a world at peace, family security*, and *freedom* the most important terminal values, and *an exciting life, pleasure, social recognition*, and *a world of beauty* the least important. Moreover, the religious, less religious, and nonreligious all agree in ranking the instrumental values *honest, ambitious,* and *responsible* highest, and *imaginative, intellectual, logical,* and *obedient* lowest in importance.

At the same time, the several groups varying in religion also differ significantly in many ways. Americans identifying themselves as Jews generally place a higher value than do Christians on *equality, pleasure, family security, inner harmony,* and *wisdom,* and on instrumental values emphasizing personal competence—being *capable, independent, intellectual,* and *logical*. The average nonbeliever's value profile is similar in many respects to that obtained for Jews. Both nonbeliever and Jew put relatively less emphasis than Christians do on such conventional values as being *clean, obedient,* and *polite.*

Two values—*salvation* and *forgiving*—stand out above all the others as the most distinctively Christian values. Whereas Jews and nonbelievers rank *salvation* last, Christians generally rank it considerably higher—third on the average for Baptists and anywhere from ninth to fourteenth for the remaining Christian groups.

REFERENCES

Beech, R. P., & Schoeppe, A. A developmental study of value systems in adolescence. Paper presented at the meeting of the American Psychological Association, Miami Beach, Fla., 1970.

Brawer, F. B. *Values and the generation gap: Junior college freshmen and faculty.* ERIC Clearinghouse for Junior Colleges, Monograph Series, No. 11, Washington, D. C.: American Association of Junior Colleges (One Dupont Circle, N.W., 20036), 1971.

Erikson, E. H. *Childhood and society.* New York: Norton, 1950.

Kohlberg, L. The development of children's orientations toward a moral order. I Sequences in the development of moral thought. *Vita Humana*, 1963, **6**, 11-33.

McLellan, D. D. Values, value systems, and the developmental structure of moral judgment. Unpublished M.A. thesis, Michigan State University Library, 1970.

Rokeach, M. Value systems in religion. *Review of Religious Research*, 1969, **11**, 3-23.(a)

Rokeach, M. Religious values and social compassion. *Review of Religious Research*, 1969, **11**, 24-38.(b)

Rokeach, M. Commentary on the commentaries. *Review of Religious Research*, 1970, **11**, 155-162.(a)

Rokeach, M. Faith, hope, and bigotry. *Psychology Today*, April 1970.(b)

TOWARD A MODERN APPROACH TO VALUES:
THE VALUING PROCESS IN THE MATURE PERSON

Carl R. Rogers

There is a great deal of concern today with the problem of values. Youth, in almost every country, is deeply uncertain of its value orientation; the values associated with various religions have lost much of their influence; sophisticated individuals in every culture seem unsure and troubled as to the goals they hold in esteem. The reasons are not far to seek. The world culture, in all its aspects, seems increasingly scientific and relativistic, and the rigid, absolute views on values which come to us from the past appear anachronistic. Even more important, perhaps, is the fact that the modern individual is assailed from every angle by divergent and contradictory value claims. It is no longer possible, as it was in the not too distant historical past, to settle comfortably into the value system of one's forbears or one's community and live out one's life without ever examining the nature and the assumptions of that system.

In this situation it is not surprising that value orientations from the past appear to be in a state of disintegration or collapse. Men question whether there are, or can be, any universal values. It is often felt that we may have lost, in our modern world, all possibility of any general or cross-cultural basis for values. One natural result of this uncertainty and confusion is that there is an increasing concern about, interest in, and a searching for, a sound or meaningful value approach which can hold its own in today's world.

I share this general concern. As with other issues the general problem faced by the culture is painfully and specifically evident in the cultural microcosm which is called the therapeutic relationship, which is my sphere of experience.

As a consequence of this experience I should like to attempt a modest theoretical approach to this whole problem. I have observed changes in the approach to values as the individual grows from infancy to adulthood. I observe further changes when, if he is fortunate, he continues to grow toward true psychological maturity. Many of these observations grow out of my experience as therapist, where I have had the mind stretching opportunity of seeing the

Reprinted from *Journal of Abnormal and Social Psychology*, 1964, Vol. 68, No. 2, 160-167. Copyright © 1964 by the American Psychological Association. Reprinted by permission of author and publisher.

ways in which individuals move toward a richer life. From these observations I believe I see some directional threads emerging which might offer a new concept of the valuing process, more tenable in the modern world. I have made a beginning by presenting some of these ideas partially in previous writings (Rogers, 1951, 1959); I would like now to voice them more clearly and more fully.

SOME DEFINITIONS

Charles Morris (1956, pp. 9-12) has made some useful distinctions in regard to values. There are "operative values," which are the behaviors of organisms in which they show preference for one object or objective rather than another. The lowly earthworm, selecting the smooth arm of a Y maze rather than the arm which is paved with sandpaper, is giving an indication of an operative value.

There are also "conceived values," the preference of an individual for a symbolized object. "Honesty is the best policy" is such a conceived value.

There is also the term "objective value," to refer to what is objectively preferable, whether or not it is sensed or conceived of as desirable. I will be concerned primarily with operative or conceptualized values.

INFANT'S WAY OF VALUING

Let me first speak about the infant. The living human being has, at the outset, a clear approach to values. We can infer from studying his behavior that he prefers those experiences which maintain, enhance, or actualize his organism, and rejects those which do not serve this end. Watch him for a bit:

> Hunger is negatively valued. His expression of this often comes through loud and clear.
> Food is positively valued. But when he is satisfied, food is negatively valued, and the same milk he responded to so eagerly is not spit out, or the breast which seemed so satisfying is now rejected as he turns his head away from the nipple with an amusing facial expression of disgust and revulsion.
> He values security, and the holding and caressing which seem to communicate security.
> He values new experience for its own sake, and we observe this in his obvious pleasure in discovering his toes, in his searching movements, in his endless curiosity.
> He shows a clear negative valuing of pain, bitter tastes, sudden loud sounds.

All of this is commonplace, but let us look at these facts in terms of what they tell us about the infant's approach to values. It is first of all a flexible, changing, valuing *process*, not a fixed system. He likes food and dislikes the same food. He values security and rest, and rejects it for new experience. What is going on seems best described as an organismic valuing process, in which each element, each moment of what he is experiencing is somehow weighed, and selected or rejected, depending on whether, at that moment, it tends to actualize the organism or not. This complicated weighing of experience is clearly an organismic, not a conscious or symbolic function. These are operative, not conceived values. But this process can nonetheless deal with complex value problems. I would remind you of the experiment in which young infants had spread in front of them a score or more of dishes of natural (that is, unflavored) foods. Over a period of time they clearly tended to value the foods which enhanced their own survival, growth, and development. If for a time a child gorged himself on starches, this would soon be balanced by a protein "binge." If at times he chose a diet deficient in some vitamin, he would later seek out foods rich in this very vitamin. The physiological wisdom of his body guided his behavioral movements, resulting in what we might think of as objectively sound value choices.

Another aspect of the infant's approach to values is that the source or locus of the evaluating process is clearly within himself. Unlike many of us, he *knows* what he likes and dislikes, and the origin of these value choices lies strictly within himself. He is the center of the valuing process, the evidence for his choices being supplied by his own senses. He is not at this point influenced by what his parents think he should prefer, or by what the church says, or by the opinion of the latest "expert" in the field, or by the persuasive talents of an advertising firm. It is from within his own experiencing that his organism is saying in nonverbal terms, "This is good for me." "That is bad for me." "I like this." "I strongly dislike that." He would laugh at our concern over values, if he could understand it.

CHANGE IN THE VALUING PROCESS

What happens to this efficient, soundly based valuing process? By what sequence of events do we exchange it for the more rigid, uncertain, inefficient approach to values which characterizes most of us as adults? Let me try to state briefly one of the major ways in which I think this happens.

The infant needs love, wants it, tends to behave in ways which will bring a repetition of this wanted experience. But this brings complications. He pulls baby sister's hair, and finds it satisfying to hear her wails and protests. He then hears that he is "a naughty, bad boy," and this may be reinforced by a slap on the hand. He is cut off from affection. As this experience is repeated, and many, many others like it, he gradually learns that what "feels good" is often "bad" in the eyes of significant others. Then the next step occurs, in which he comes to take the same attitude toward himself which these others have taken. Now, as he pulls his sister's hair, he solemnly intones, "Bad, bad boy." He is introjecting the value judgment of another, taking it in as his own. To that degree he loses touch with his own organismic valuing process. He has deserted the wisdom of his organism, giving up the locus of evaluation, and is trying to behave in terms of values set by another, in order to hold love.

Or take another example at an older level. A boy senses, though perhaps not consciously, that he is more loved and prized by his parents when he thinks of being a doctor than when he thinks of being an artist. Gradually he introjects the values attached to being a doctor. He comes to want, above all, to be a doctor. Then in college he is baffled by the fact that he repeatedly fails in chemistry, which is absolutely necessary to becoming a physician, in spite of the fact that the guidance counselor assures him he has the ability to pass the course. Only in counseling interviews does he begin to realize how completely he has lost touch with his organismic reactions, how out of touch he is with his own valuing process.

Perhaps these illustrations will indicate that in an attempt to gain or hold love, approval, esteem, the individual relinquishes the locus of evaluation which was his in infancy, and places it in others. He learns to have a basic *dis*trust for his own experiencing as a guide to his behavior. He learns from others a large number of conceived values, and adopts them as his own, even though they may be widely discrepant from what he is experiencing.

SOME INTROJECTED PATTERNS

It is in this fashion, I believe, that most of us accumulate the introjected value patterns by which we live. In the fantastically complex culture of today, the patterns we introject as desirable or undesirable come from a variety of sources and are often highly contradictory. Let me list a few of the introjections which are commonly held.

> Sexual desires and behaviors are mostly bad. The sources of this construct are many—parents, church, teachers.
>
> Disobedience is bad. Here parents and teachers combine with the military to emphasize this concept. To obey is good. To obey without question is even better.
>
> Making money is the highest good. The sources of this conceived value are too numerous to mention.
>
> Learning an accumulation of scholarly facts is highly desirable. Education is the source.
>
> Communism is utterly bad. Here the government is a major source.
>
> To love thy neighbor is the highest good. This concept comes from the church, perhaps from the parents.
>
> Cooperation and teamwork are preferable to acting alone. Here companions are an important source.
>
> Cheating is clever and desirable. The peer group again is the origin.
>
> Coca-Colas, chewing gum, electric refrigerators, and automobiles are all utterly desirable. From Jamaica to Japan, from Copenhagen to Kowloon, the "Coca-Cola culture" has come to be regarded as the acme of desirability.

This is a small and diversified sample of the myriads of conceived values which individuals often introject, and hold as their own, without ever having considered their inner organismic reactions to these patterns and objects.

COMMON CHARACTERISTICS OF ADULT VALUING

I believe it will be clear from the foregoing that the usual adult—I feel I am speaking for most of us—has an approach to values which has these characteristics:

> The majority of his values are introjected from other individuals or groups significant to him, but are regarded by him as his own.
>
> The source or locus of evaluation on most matters lies outside of himself.
>
> The criterion by which his values are set is the degree to which they will cause him to be loved, accepted, or esteemed.

These conceived preferences are either not related at all, or not clearly related, to his own process of experiencing.

Often there is a wide and unrecognized discrepancy between the evidence supplied by his own experience, and these conceived values.

Because these conceptions are not open to testing in experience, he must hold them in a rigid and unchanging fashion. The alternative would be a collapse of his values. Hence his values are "right."

Because they are untestable, there is no ready way of solving contradictions. If he has taken in from the community the conception that money is the *summum bonum* and from the church the conception that love of one's neighbor is the highest value, he has no way of discovering which has more value for *him*. Hence a common aspect of modern life is living with absolutely contradictory values. We calmly discuss the possibility of dropping a hydrogen bomb on Russia, but find tears in our eyes when we see headlines about the suffering of one small child.

Because he has relinquished the locus of evaluation to others, and has lost touch with his own valuing process, he feels profoundly insecure and easily threatened in his values. If some of these conceptions were destroyed, what would take their place? This threatening possibility makes him hold his value conceptions more rigidly or more confusedly, or both.

FUNDAMENTAL DISCREPANCY

I believe that this picture of the individual, with values mostly introjected, held as fixed concepts, rarely examined or tested, is the picture of most of us. By taking over the conceptions of others as our own, we lose contact with the potential wisdom of our own functioning, and lose confidence in ourselves. Since these value constructs are often sharply at variance with what is going on in our own experiencing, we have in a very basic way divorced ourselves from ourselves, and this accounts for much of modern strain and insecurity. This fundamental discrepancy between the individual's concept and what he is actually experiencing, between the intellectual structure of his values and the valuing process going on unrecognized within—this is a part of the fundamental estrangement of modern man from himself.

RESTORING CONTACT WITH EXPERIENCE

Some individuals are fortunate in going beyond the picture I have just given, developing further in the direction of psychological maturity. We see this happen in psychotherapy where we endeavor to provide a climate favorable to the growth of the person. We also see it happen in life, whenever life provides a therapeutic climate for the individual. Let me concentrate on this further maturing of a value approach as I have seen it in therapy.

As the client senses and realizes that he is prized as a person[1] he can slowly begin to value the different aspects of himself. Most importantly, he can begin, with much difficulty at first, to sense and to feel what is going on within him, what he is feeling, what he is experiencing, how he is reacting. He uses his experiencing as a direct referent to which he can turn in forming accurate

---

[1]The therapeutic relationship is not devoid of values. When it is most effective it is, I believe, marked by one primary value, namely, that this person (the client) has *worth*.

conceptualizations and as a guide to his behavior. Gendlin (1961, 1962) has elaborated the way in which this occurs. As his experiencing becomes more and more open to him, as he is able to live more freely in the process of his feelings, then significant changes begin to occur in his approach to values. It begins to assume many of the characteristics it had in infancy.

INTROJECTED VALUES IN RELATION TO EXPERIENCING

Perhaps I can indicate this by reviewing a few of the brief examples of introjected values which I have given, and suggesting what happens to them as the individual comes closer to what is going on within him.

> The individual in therapy looks back and realizes, "But I *enjoyed* pulling my sister's hair—and that doesn't make me a bad person."
>
> The student failing chemistry realizes, as he gets close to his own experiencing, "I don't like chemistry; I don't value being a doctor, even though my parents do; and I am not a failure for having these feelings."
>
> The adult recognizes that sexual desires and behavior may be richly satisfying and permanently enriching in their consequences, or shallow and temporary and less than satisfying. He goes by his own experiencing, which does not always coincide with social norms.
>
> He recognizes freely that this communist book or person expresses attitudes and goals which he shares as well as ideas and values which he does not share.
>
> He realizes that at times he experiences cooperation as meaningful and valuable to him, and that at other times he wishes to be alone and act alone.

VALUING IN THE MATURE PERSON

The valuing process which seems to develop in this more mature person is in some ways very much like that in the infant, and in some ways quite different. It is fluid, flexible, based on this particular moment, and the degree to which this moment is experienced as enhancing and actualizing. Values are not held rigidly, but are continually changing. The painting which last year seemed meaningful now appears uninteresting, the way of working with individuals which was formerly experienced as good now seems inadequate, the belief which then seemed true is now experienced as only partly true, or perhaps false.

Another characteristic of the way this person values experience is that it is highly differentiated, or as the semanticists would say, extensional. The examples in the preceding section indicate that what were previously rather solid monolithic introjected values now become differentiated, tied to a particular time and experience.

Another characteristic of the mature individual's approach is that the locus of evaluation is again established firmly within the person. It is his own experience which provides the value information or feedback. This does not mean that he is not open to all the evidence he can obtain from other sources. But it means that this is taken for what it is—outside evidence—and is not as significant as his own reactions. Thus he may be told by a friend that a new book is very disappointing. He reads two unfavorable reviews of the book. Thus his tentative hypothesis is that he will not value the book. Yet if he reads the book his valuing will be based upon the reactions it stirs in *him*, not on what he has been told by others.

There is also involved in this valuing process a letting oneself down into the immediacy of what one is experiencing, endeavoring to sense and to clarify all its complex meanings. I think of a client who, toward the close of therapy, when puzzled about an issue, would put his head in his hands and say, "Now what *is* it that I'm feeling? I want to get next to it. I want to learn what it is." Then he would wait, quietly and patiently, trying to listen to himself, until he could discern the exact flavor of the feelings he was experiencing. He, like others, was trying to get close to himself.

In getting close to what is going on within himself, the process is much more complex than it is in the infant. In the mature person it has much more scope and sweep. For there is involved in the present moment of experiencing the memory traces of all the relevant learnings from the past. This moment has not only its immediate sensory impact, but it has meaning growing out of similar experiences in the past (Gendlin, 1962). It has both the new and the old in it. So when I experience a painting or a person, my experiencing contains within it the learnings I have accumulated from past meetings with paintings or persons, as well as the new impact of this particular encounter. Likewise the moment of experiencing contains, for the mature adult, hypotheses about consequences. "It is not pleasant to express forthrightly my negative feelings to this person, but past experience indicates that in a continuing relationship it will be helpful in the long run." Past and future are both in this moment and enter into the valuing.

I find that in the person I am speaking of (and here again we see a similarity to the infant), the criterion of the valuing process is the degree to which the object of the experience actualizes the individual himself. Does it make him a richer, more complete, more fully developed person? This may sound as though it were a selfish or unsocial criterion, but it does not prove to be so, since deep and helpful relationships with others are experienced as actualizing.

Like the infant, too, the psychologically mature adult trusts and uses the wisdom of his organism, with the difference that he is able to do so knowingly. He realizes that if he can trust all of himself, his feelings and his intuitions may be wiser than his mind, that as a total person he can be more sensitive and accurate than his thoughts alone. Hence he is not afraid to say, "I feel that this experience [or this thing, or this direction] is good. Later I will probably know *why* I feel it is good." He trusts the totality of himself, having moved toward becoming what Lancelot Whyte (1950) regards as "the unitary man."

It should be evident from what I have been saying that this valuing process in the mature individual is not an easy or simple thing. The process is complex, the choices often very perplexing and difficult, and there is no guarantee that the choice which is made will in fact prove to be self-actualizing. But because whatever evidence exists is available to the individual, and because he is open to his experiencing, errors are correctable. If this chosen course of action is not self-enhancing this will be sensed and he can make an adjustment or revision. He thrives on a maximum feedback interchange, and thus, like the gyroscopic compass on a ship, can continually correct his course toward his true goal of self-fulfillment.

SOME PROPOSITIONS REGARDING THE VALUING PROCESS

Let me sharpen the meaning of what I have been saying by stating two propositions which contain the essential elements of this viewpoint. While it may not be possible to devise empirical tests of each proposition in its entirety, yet each is to some degree capable of being tested through the methods of psychological science. I would also state that though the following propositions are stated firmly in order to give them clarity, I am actually advancing them as decidedly tentative hypotheses.

Hypothesis I. There is an organismic base for an organized valuing process within the human individual.

It is hypothesized that this base is something the human being shares with the rest of the animate world. It is part of the functioning life process of any healthy organism. It is the capacity for receiving feedback information which enables the organism continually to adjust its behavior and reactions so as to achieve the maximum possible self-enhancement.

Hypothesis II. This valuing process in the human being is effective in achieving self-enhancement to the degree that the individual is open to the experiencing which is going on within himself.

I have tried to give two examples of individuals who are close to their own experiencing: the tiny infant who has not yet learned to deny in his awareness the processes going on within; and the psychologically mature person who has relearned the advantages of this open state.

There is a corollary to this second proposition which might be put in the following terms. One way of assisting the individual to move toward openness to experience is through a relationship in which he is prized as a separate person, in which the experiencing going on within him is empathically understood and valued, and in which he is given the freedom to experience his own feelings and those of others without being threatened in doing so.

This corollary obviously grows out of therapeutic experience. It is a brief statement of the essential qualities in the therapeutic relationship. There are already some empirical studies, of which the one by Barrett-Lennard (1962) is a good example, which give support to such a statement.

PROPOSITIONS REGARDING THE OUTCOMES OF THE VALUING PROCESS

I come now to the nub of any theory of values or valuing. What are its consequences? I should like to move into this new ground by stating bluntly two propositions as to the qualities of behavior which emerge from this valuing process. I shall then give some of the evidence from my experience as a therapist in support of these propositions.

Hypothesis III. In persons who are moving toward greater openness to their experiencing, there is an organismic commonality of value directions.

Hypothesis IV. These common value directions are of such kinds as to enhance the development of the individual himself, of others in his community, and to make for the survival and evolution of his species.

It has been a striking fact of my experience that in therapy, where individuals are valued, where there is greater freedom to feel and to be, certain value directions seem to emerge. These are not chaotic directions but instead exhibit a surprising commonality. This commonality is not dependent on the personality of the therapist, for I have seen these trends emerge in the clients of therapists sharply different in personality. This commonality does not seem to be due to the influences of any one culture, for I have found evidence of these directions in cultures as divergent as those of the United States, Holland, France, and Japan. I like to think that this commonality of value directions is due to the fact that we all belong to the same species—that just as a human infant tends, individually, to select a diet similar to that selected by other human infants, so a client in therapy tends, individually, to choose value directions similar to those chosen by other clients. As a species there may be certain elements of experience which tend to make for inner development and which would be chosen by all individuals if they were genuinely free to choose.

Let me indicate a few of these value directions as I see them in my clients as they move in the direction of personal growth and maturity.

They tend to move away from façades. Pretense, defensiveness, putting up a front, tend to be negatively valued.

They tend to move away from "oughts." The compelling feeling of "I ought to do or be thus and so" is negatively valued. The client moves away from being what he "ought to be," no matter who has set that imperative.

They tend to move away from meeting the expectations of others. Pleasing others, as a goal in itself, is negatively valued.

Being real is positively valued. The client tends to move toward being himself, being his real feelings, being what he is. This seems to be a very deep preference.

Self-direction is positively valued. The client discovers an increasing pride and confidence in making his own choices, guiding his own life.

One's self, one's own feelings come to be positively valued. From a point where he looks upon himself with contempt and despair, the client comes to value himself and his reactions as being of worth.

Being a process is positively valued. From desiring some fixed goal, clients come to prefer the excitement of being a process of potentialities being born.

Sensitivity to others and acceptance of others is positively valued. The client comes to appreciate others for what they are, just as he has come to appreciate himself for what he is.

Deep relationships are positively valued. To achieve a close, intimate, real, fully communicative relationship with another person seems to meet a deep need in every individual, and is very highly valued.

Perhaps more than all else, the client comes to value an openness to all of his inner and outer experience. To be open to and sensitive to his own *inner* reactions and feelings, the reactions and feelings of others, and the realities of the objective world—this is a direction which he clearly prefers. This openness becomes the client's most valued resource.

These then are some of the preferred directions which I have observed in individuals moving toward personal maturity. Though I am sure that the list I have given is inadequate and perhaps to some degree inaccurate, it holds for me exciting possibilities. Let me try to explain why.

I find it significant that when individuals are prized as persons, the values they select do not run the full gamut of possibilities. I do not find, in such a climate of freedom, that one person comes to value fraud and murder and thievery, while another values a life of self-sacrifice, and another values only money. Instead there seems to be a deep and underlying thread of commonality. I believe that when the human being is inwardly free to choose whatever he deeply values, he tends to value those objects, experiences, and goals which make for his own survival, growth, and development, and for the survival and development of others. I hypothesize that it is *characteristic* of the human organism to prefer such actualizing and socialized goals when he is exposed to a growth promoting climate.

A corollary of what I have been saying is that in *any* culture, given a climate of respect and freedom in which he is valued as a person, the mature individual would tend to choose and prefer these same value directions. This is a significant hypothesis which could be tested. It means that though the individual of whom I am speaking would not have a consistent or even a stable system of conceived values, the valuing process within him would lead to emerging value directions which would be constant across cultures and across time.

Another implication I see is that individuals who exhibit the fluid valuing process I have tried to describe, whose value directions are generally those I have listed, would be highly effective in the ongoing process of human evolution. If the human species is to survive at all on this globe, the human being must become more readily adaptive to new problems and situations, must be able to select that which is valuable for development and survival out of new and complex situations, must be accurate in his appreciation of reality if he is to make such selections. The psychologically mature person as I have described him has, I believe, the qualities which would cause him to value those experiences which would make for the survival and enhancement of the human race. He would be a worthy participant and guide in the process of human evolution.

Finally, it appears that we have returned to the issue of universality of values, but by a different route. Instead of universal values "out there," or a universal value system imposed by some group—philosophers, rulers, priests, or psychologists—we have the possibility of universal human value directions *emerging* from the experiencing of the human organism. Evidence from therapy indicates that both personal and social values emerge as natural, and experienced, when the individual is close to his own organismic valuing process. The suggestion is that though modern man no longer trusts religion or science or philosophy nor any system of beliefs to *give* him values, he may find an organismic valuing base within himself which, if he can learn again to be in touch with it, will prove to be an organized, adaptive, and social approach to the perplexing value issues which face all of us.

REFERENCES

BARRETT-LENNARD, G. T. Dimensions of therapist response as casual factors in therapeutic change. *Psychol. Monogr.*, 1962, 76, (43, Whole No. 562).

GENDLIN, E. T. Experiencing: A variable in the process of therapeutic change. *Amer. J. Psychother.*, 1961, 15, 233-245.

GENDLIN, E. T. *Experiencing and the creation of meaning.* Glencoe, Ill.: Free Press, 1962.

MORRIS, C. W. *Varieties of human value.* Chicago: Univer. Chicago Press, 1956.

ROGERS, C. R. *Client-centered therapy.* Boston: Houghton Mifflin, 1951.

ROGERS, C. R. A theory of therapy, personality and interpersonal relationships. In S. Koch (Ed.), *Psychology: A study of a science.* Vol. 3. *Formulations of the person and the social context.* New York: McGraw-Hill, 1959. Pp. 185-256.

WHYTE, L. L. *The next development in man.* New York: Mentor Books, 1950.

BASIC PROBLEMS IN AXIOLOGY

Risieri Frondizi

Value Problems in Daily Life

Fundamental value problems are not only a concern of philosophers; they are also present in our daily life. There is not a discussion or disagreement about a person's behavior, a woman's elegance, or the enjoyment of a meal, that does not have as its basis a question of values. The most complicated axiological problems are debated daily in the street, in parliament, and in the most modest homes, although with an attitude and in a language which can hardly be called philosophical. Nevertheless, the discussions generally reveal the two extreme positions of axiology. When two people disagree whether a meal or a drink is pleasant, or not, each fails to change the other's opinion, one of them usually ends the discussion by saying that he likes it, and no one can convince him otherwise. If it is a discussion between educated people, then someone may recall the Latin adage: *de gustibus non est disputandum.*

This proverb can put an end to an argument, either of the ordinary or more sophisticated variety, but it does not solve the basic problem which underlies such discussion. Is it true that one cannot argue about taste? Is it improper, therefore, to speak of people of bad taste? Have there not been debates for years about the esthetic value of a considerable number of statues, paintings and poems? Are these discussions, then, futile, and is there no way of determining the value of an artistic work or the conduct of a person?

He who supports the thesis *de gustibus non disputandum* wishes to affirm a peculiar characteristic of value, i.e., the intimate and immediate nature of valuation. The pleasure produced in us by a glass of good wine, the reading of a poem, a prelude by Chopin, is something personal, intimate, private, and frequently, ineffable. We do not wish to relinquish this intimacy, for if we did, an essential part of esthetic enjoyment would then slip through our fingers. How can anyone convince us with syllogisms and learned quotations, when our pleasure is so immediate and direct that it does not admit any possibility of error?

However, if one does not take refuge in subjectivity, and tries to keep a cool head even though his heart is anything but calm, he will soon find that this doctrine is not satisfactory. What would become of ethical norms and esthetic masterpieces if each of us abided by his own particular way of looking

---

From *What Is Value?* by Risieri Frondizi. La Salle, Ill., Open Court Publishing Company, 1971, pp. 17-34.

at things? How can chaos be avoided, unless there are standards of value, norms of behavior? If every one carries his own yardstick of valuation, by what standard shall we decide axiological conflicts? Esthetic and moral education would be impossible; moral life will be meaningless; repentance of sin would seem absurd. "Moral" for whom? "Sin" for whom? one would have to ask constantly. On the other hand, if one were to measure esthetic value by the intensity of individual or collective emotion, greater value would accrue to screen or radio melodrama (which has such wide sentimental appeal) than to *Hamlet* or *King Lear*, which have appealed to a much smaller audience. If we make man the measure of esthetic value and moral law, it would appear that there would be, strictly speaking, neither "good taste" nor morality.

This issue is an old one. Indeed, it is as old as axiology itself, and the history of value theory could be written around this basic problem and the various solutions which have been proposed to solve it.

ARE VALUES OBJECTIVE OR SUBJECTIVE?

While it is not easy to reduce to simple terms the constellation of problems with which axiology is concerned today, the core of the problem may be summed up in the following question: *Are things valuable because we desire them, or do we desire them because they are valuable?* Does desire, pleasure or interest give value to an object, or are we interested because such objects possess a value which is prior and foreign to our psychological and organic reactions? Though in a different context, we find the two attitudes in Shakespeare's *Troilus and Cressida*: (II, 2):

> *Hector:* Brother, she is not worth what she doth cost the holding.
> *Troilus:* What is aught, but as 'tis valued?
> *Hector:* But value dwells not in particular will;
> It holds his estimate and dignity
> As well wherein 'tis precious of itself.

If one prefers a more technical and traditional way of posing the problem, one may ask: Are values objective or subjective?

The question requires prior clarification of terms in order to prevent us from falling into a *disputatio de nominem*. Value is "objective" if its existence and nature is independent of a subject; conversely, it is "subjective" if it owes its existence, its sense, or its validity, to the feelings or attitudes of the subject.

An illustration can further clarify the sense of the problem. As it has been pointed out, physical objects have certain qualities, called "primary," which are inherent in the objects themselves and others, such as sense or "secondary" qualities, which depend at least partially upon a subject who perceives them. To take a specific value: which of these two types of qualities is beauty closest to? Is it like length, which does not depend on the subject? Or is it instead like smell, which in order to exist requires the presence of a subject to perceive it, since an odor which no one can smell is nonexistent?

At times we lean toward subjectivism and we think we have discovered in the contrary point of view a mere delusion, similar to that suffered by the victim of hallucinations who is frightened by the phantoms created by his own imagination. On the other hand, there are times when it appears evident to us that values are objective realities to which we should submit, since they possess an overpowering force which brushes aside our preferences and overcomes our will. Do we not, at times, make efforts to create a work of art—a poem, a painting, a novel—only to give it up as a failure immediately we note that beauty is lacking in our creation? Something similar occurs when we appraise positively objects which we do not like, or when we notice the scant value of that which arouses us because of purely personal reasons.

But, returning to the first position, what values would objects have if we passed them by indifferently, if they did not cause us enjoyment or satisfaction, if we did not desire them, or were unable to desire them?

One point seems clear: we cannot speak of values without considering actual or possible valuations. In fact, what sense would values have if they could completely escape man's appreciation? How would we know that such values exist, if they existed outside the sphere of human valuation? In this point, subjectivism seems to be on firm ground; value cannot be divorced from valuation. Objectivism creates a basic distinction here which prevents us from pursuing the already open road of subjectivity. It is true, claims the objectivist, that valuation is subjective, but a distinction should be made between valuation and value. Value is prior to valuation. If there were no values, what would we evaluate? To confuse valuation with value is like confusing perception with the object perceived. Perception does not create the object; it grasps it. The same thing happens in the case of valuation. What is subjective is the apprehension of values, but values exist before being apprehended.

To show the weakness of this kind of reasoning, subjectivism appeals to experience. If values were objective, it asserts, then individuals would have come to an agreement about values, and value objects. But this is not the case; we find disagreement everywhere.

"Is there agreement concerning the basic principles of science?" retorts the objectivist. The mistakes made by certain persons do not invalidate the objectivity of truth. There are still people who believe in spontaneous generation and evil spirits. Truth does not depend on the opinion of individuals, but on the objectivity of facts; hence it cannot be strengthened nor weakened by the democratic procedure of counting votes. Similarly, in the case of values. The opinion of those of poor taste does not impair the beauty of a work of art. It would be idle to try to obtain unanimity of opinion. "But," the objectivist goes on to say, "there is still another point: the discrepancy refers to value objects, not to values. No one can fail to appreciate beauty; what may happen is that people may not recognize the presence of beauty in a certain object, whether this be a statue, a painting, or a symphony. Similarly, in the case of the other values: who can fail to appraise utility, prefer the pleasant, or appreciate honesty?"

"It is not so," the subjectivist will probably reply; "the disagreement referred to is extended to values themselves. When an Italian and an American disagree about the elegance of a pair of shoes, such an argument about a concrete object is due to a different manner of understanding elegance itself.

Though it is true that sometimes we disagree about value objects, such disagreement frequently reveals a profound discrepancy between values such as beauty, justice or elegance."

"There are concrete instances," the subjectivist will go on to say, "which show clearly the subjectivity of values. Postage stamps constitute one such case. Where does the value of used stamps lie? Is there something in the quality of the paper, or in the beauty of the drawing, or in the print, which explains the value they have? They would have no value, were it not for the philatelists. Our desire to collect them is what has bestowed value upon them. If this interest is lost, the value which has been conferred upon them disappears *ipso facto*. Although the problem is somewhat more complex, something similar occurs in the case of esthetic values. They, too, depend on a series of conditions—subjective, cultural and sociological. What esthetic value would a painting have if men did not have eyes? And what sense would there be in talking about the esthetic value of music if God had condemned us to eternal deafness? In the last analysis, we value what we desire and what pleases us."

"Not so," replies the objectivist; "we value also that which displeases us. Who likes to risk his life to save a man who is drowning, especially if that man is our enemy? Nevertheless, we do it because it is the right thing to do. We place what is right above our pleasure or desire. Duty is objective and is based on a moral value which is equally so, and lies beyond the fluctuations of our likes and dislikes, interests, and desires. Or, to refer to a more ordinary type of illustration: who likes the dentist's "torture"? Nevertheless, we value his work. Is it pleasant to have a leg amputated? Displeasure notwithstanding, we are grateful to the man who amputates our leg when this would save our life. One must distinguish between valuation as a psychological act, and the truth of the valuation. As an experience, a wrong perception is as much a perception as a true one; yet we do not, on that account, equate the two when we judge the accuracy involved."

Such arguments are, for the subjectivists, a sample of the superficiality of the analysis of their thesis. At first glance, it seems evident that the dentist is the cause of our pain, when he drills one of our teeth, and that consequently, the value which we ascribe to his work has nothing to do with pleasure, but is rather dictated by a higher concern. But the latter is also based on pleasure: we prefer a temporary pain for a few minutes to a prolonged toothache from a cavity. Or, if it is a matter of esthetic motivation that makes us willing to submit to torture in the dentist's chair (this, especially in the case of women) it is because we prefer the more lasting pleasure afforded by pleasant-looking teeth to the uncomfortable feeling brought on by the necessity of having to exhibit a sickly-looking set of teeth. The example of the leg amputation reveals even more sharply the confusion which we have indicated. We accept the pain of the leg amputation because it saves us from a greater pain. In both cases, we sacrifice temporary pleasure for a lasting one.

"One cannot formulate a theory based on two examples," argues the objectivist. "What satisfaction is derived by us through the act of saving the life of our enemy? Perhaps it may be argued that it is the satisfaction of having performed our duty. Our duty cannot be identified with what is pleasant; if this were so, everybody would perform his duty. Honesty depends on our capacity to overcome the claims of our pleasures, appetites and comforts. Pleasure operates on a low level of our personality, and we cannot sacrifice the highest (which is what moral values are) to the lowest. But even within the realm of the pleasant, it is necessary to distinguish between what pleases us and what we recognize as being pleasant. We frequently differentiate between what is pleasant and what we like because of personal or circumstantial motives. I still like to listen to an old waltz which used to move me when I was an adolescent; yet, I do not admit that it is more pleasant than, for example, Schubert's *Unfinished Symphony*. Similarly, in the case of desire, it is necessary to separate what is desired from what is desirable. The fact that people desire something does not change it to something desirable. By the same token, I may not at the moment have the faintest desire to drink champagne, but I cannot fail to admit that it is a pleasant, desirable drink.

The subjectivist does not believe that one should postulate a world of the pleasant and desirable per se; both are related to real, specific pleasures and desires. When I admit that something is pleasant which, under different circumstances, I find unpleasant, this is not because I recognize an intrinsic quality, foreign to concrete experiences of pleasure. For example, if I recognize that champagne is pleasant, although it is unpleasant to drink it for breakfast, it is because I consider that on other occasions I like it. I am opposing two personal reactions: one transitory—which is the present—and another, more permanent; and not my personal reaction to the supposed objectivity of the value known as "pleasant." Anything that is pleasant in an object is derived from the pleasure which it calls forth. Could anything be pleasant if it did not please anybody, or if there were no possibility that it might please? The pleasant is a concept which is based on personal experiences of pleasure, and does not exist in a metaphysical world. If we sever the connections between pleasure and what is pleasurable, the latter disappears completely. Similar considerations would have to be granted in the case of the desired and the desirable. When we define the desirable as that which is worthy of being desired, we do not transfer the concept to a meta-empirical world; what we mean is that it would be desired by a person in "normal" circumstances. The example involving the postage stamps proves clearly that real and actual desire is what confers value upon things; when this is lacking, value disappears.

The subjectivist arguments do not succeed in convincing those who adhere to the objectivist thesis. The latter maintain that an entire axiological theory cannot be made to rest on an example involving postage stamps; an examination of any other case proves the opposite. "Hence," they repeat, "things do not have value because we desire them, but we desire them precisely because they are valuable. In fact, it would appear that we do not desire them out of sheer caprice, or without reason, but because something is within them which makes them *desirable* in the two senses of being capable and worthy of being desired."

SUGGESTIONS FOR A NEW WAY OF LOOKING
AT THE PROBLEM

Though the mind is enriched by the arguments offered by both sides, the discussion shows no sign of abating. Nor are the problems settled by deciding in favor of one or the other position. If we admit that value is of a subjective nature, there still remains to be decided just which aspect of subjectivity it is that gives value its force. Do objects possess value because they please us? Or perhaps because we desire them or we are interested in them? And why do we have an interest in certain things and not in others? Why do we like or prefer this to that? Is it an arbitrary psychological reaction or is there something in the object which compels us to react in a certain manner?

And so we land in the realm of objectivism. Nor are things crystal clear here. Is value completely alien to the biological and psychological constitution of man? Or is it true that all objectivity consists in the fact that man cannot fail to recognize value, once he is confronted by it? Isn't objectivity, possibly, of a completely different order? What about social objectivity, for example, in which the objectivity is based on the intersubjective character of the reaction? And once again, we are back to subjectivism.

This going around in circles, from one position to the opposite one, and then back again to the first, makes us think that perhaps the difficulty arises because the problem has been poorly stated. Does value necessarily have to be objective or subjective? Aren't we perhaps confused by our eagerness to reduce the whole to one of its essential elements? It is possible, for example, that pleasure, desire or interest, are a necessary but not a sufficient condition, and that they do not exclude objective elements. This is to say that value may be the result of a tension between the subject and the object, and therefore presents a subjective as well as an objective aspect, deceiving those who look only at one side of the coin.

At the moment, let us attempt other paths. Do all values have identical character? The central problem concerns the nature of value. Should we not, before attempting to determine this, ask whether all values have a similar nature as concerns objectivity or subjectivity? Will not the element of subjectivity or objectivity vary according to the type of value? Let us explore this possibility for a moment, examining values which may belong to different hierarchies.

Let us begin with the lowest: those which pertain to pleasure or displeasure. I drink a glass of wine and find it pleasant. Where is the pleasant quality—in me or in the wine? Are we faced with a subjective or an objective value? It would seem that the pleasantness is a quality possessed by the wine, since Coca-Cola, for example, does not have the same pleasant effect. If I think a moment, I notice, nevertheless, that another person might be able to assert the exact opposite: that he likes Coca-Cola, and dislikes the wine. If this is so, it is not the object, but the subject that is the source of value. If everyone reacts differently in the presence of the same stimulus, then the difference is probably traceable to the subject. It is not an acceptable refutation to say that there are people of poor taste who are incapable of enjoying wine, or whose taste has become perverted so that they find most pleasant what is not so. If we compare French and Italian wines, both of recognized quality, we notice that preferences are probably due to personal idiosyncrasies, or to habits acquired, from living in one country or the other. This is where the proverb *de gustibus non disputandum* makes sense, since it is a recognition of the predominance of the subjective over the objective on the lowest axiological level.

This predominance will be lost if we jump to the highest point on the axiological scale: to ethical values, for example. Is an attitude which we judge to be honest or dishonest, is a verdict just or unjust, dependent upon our feeling at the time? Of course not. We have to be above those subjective conditions which distort our ethical evaluation. What sort of judge would allow his verdicts to be affected by a stomach ache, or by the quarrels he might have had with his wife? Ethical value is so forceful that it compels us to acknowledge it, even against our personal desires and interests. At least, it appears that the element of objectivity is, in this sense, much greater than when we deal with what is pleasant.

Between these two extremes one finds other values: the useful, the vital, the esthetic. It is in the last group that the balance between the subjective and the objective appears greatest, although also varying according to the nature of the esthetic value. There is, for example, a predominance of the subjective element in evaluating an elegant dress—connected with fashion and other changing factors—which can be ignored when we evaluate the beauty of a painting.

THE METHODOLOGICAL PROBLEM

In recent years the impression has been gaining ground that the problem of the ultimate nature of value has entered an *impasse*. The history of science and philosophy has many times found itself face to face with a similar situation in which the main problem had to be postponed in order to give consideration to a prior problem. At the beginning of the 16th century, it was more important to find the means which would permit the discovery of new truths than to find the truths themselves. This was the contribution of Francis Bacon and Descartes, among others. Something similar occurred at the end of the 17th century when Locke put aside metaphysical questions in order first to pose the problem of the origin of our ideas; or a century later, when Kant made the nature of knowledge the central concern of theoretical philosophy to the detriment of the metaphysical problem.

In the face of the impossibility of settling the dispute between subjectivists and objectivists, many have thought that the moment has finally arrived when that question could be put aside in order to give priority to problems of method and criteria. What criteria shall we utilize to decide who is right? What is the most appropriate method of discovering the ultimate nature of value? John Dewey is one of the thinkers who believes that the main problem today is methodological. After having concerned himself with axiological questions for several decades, Dewey, at the age of 90, wrote: "In the present situation, as concerns the problem of values, the decisive question is of a methodological type." And Dewey is not alone; there are many who feel as he does, namely, that axiology will not emerge from the condition in which it finds itself, unless the problem of method is first clarified.

It is true that the method which is selected cannot be completely separated from the theoretical preferences, so that a course of action is already indicated in the statement of the problem; but it is no less true that if the criterion to be utilized is not determined with a certain degree of clarity, then the discussion is not only interminable but fruitless. On the other hand, an adequate method can shed a good deal of light on the problem, especially if the method does not carry with it an advance commitment to a definite theory.

Which is the road, then, to be followed? Two chief possibilities open up before us: one is empirical, the other a priori. Will we have to adjust to experience and abide by its decisions, or should we trust emotional intuition, as Scheler would prefer—capable of transporting us to the very intimacy of essence and assuring us of unquestionable knowledge?

Experience is the supreme judge in matters of fact; it will tell us, if we wish to undertake a complicated research problem, what people really prefer, what they really value, and what they dislike. But on the basis of the observation that people value in a certain manner, we cannot infer by way of conclusion that this is the way in which they should value. We have already seen that if we made value dependent upon facts, there would be no possibility for moral reform, since moral law would be identified with the mores of a given community. The essence of the moral reformer and of the creator in the field of art lies in not adjusting to the predominant norms or tastes, but unfurling the flag of what "ought to be" over and above people's preferences.

Are we left, then, only with the other alternative, that of infallible intuition, which asserts haughtily and not very philosophically that those who disagree with its theories have "value blindness"? What should we do if the infallible intuitions of two of these "chosen few" do not agree? And what should we think of the infallibility of intuition when it is the same individual—as is the case with Scheler himself—who, in the course of his life, experienced "infallible" intuitions which are contradictory?

These difficulties reveal to us a characteristic peculiar to philosophy. Scientific propositions, (however difficult) are judged true or false according to commonly agreed-upon criteria. One can stand firmly on this foundation upon which science rests. On the other hand, in philosophy, the criterion to be utilized, the yardstick, is also a problem under discussion. There is no yardstick to measure the yardstick. This should not cast us into the depths of despair or scepticism; it should reveal to us the complexity of philosophic problems, and make us cautiously alert to over-simplified solutions which settle problems by pushing them aside. The philosophical attitude is basically problem oriented. He who is not capable of grasping the sense of problems and who prefers to seize upon the first solution which presents itself, and which offers him illusory stability, runs the risk of being drowned in a sea of difficulties. It is because no axiological theory can be understood without first grasping the essence of the problems which that theory endeavors to solve, that we have devoted this chapter to problems of present-day value theory. But this is not the end of the dilemma. The problems indicated are the most important, but they are not the only ones. Before passing on to the proposed solutions, it would be best to glance at some other axiological problems. Since it is impossible to present them all, we shall single out those which appear to have the greatest significance.

HOW DO WE APPREHEND VALUES?

Let us limit the methodological problem to the smaller, yet no less important question of the apprehension of values. How do we grasp values?

We have seen . . . that values are not self-sustaining, but that they lead a parasitical existence; they always appear to us, resting on some carrier, or value object. This carrier is a real object—stone, canvas, paper, gesture, movement—and we perceive it through the senses. Do we perceive in the same way the value which rests on it? Let us not confuse the question: it is evident that if we do not perceive the object via the senses, in which the value is embodied, the value will be concealed from us. The question which we pose is different. We want to know whether it is through the senses or via any other means that we perceive the values once embodied. Thus, for example, when we see two apples, we perceive each one with our eyes; the similarity, however, is perceived not with the eyes of our face, but with those of our mind. It is evident that it would not have been possible to perceive the similarity intellectually, if we had not first seen the objects. This truth does not exclude the former. The same thing happens in the case of values: we can, and we should, separate the perception of real objects which serve as a vehicle for values, from the values themselves, and ask ourselves if both are perceived in a similar manner.

Aside from the interest which the problem of the perception of values offers by itself, its solution will shed light on the nature of the values themselves. Since we cannot ourselves penetrate the very being of objects "in themselves," by eliminating our own person, we ought to resign ourselves to discovering the nature of objects according to the relationship we bear to them. Thus, for example, the difference that exists between *a* horse, *the* horse (as species or essence), and a centaur, is a consequence of what we can do with them. We can see, lasso, ride *a* horse; we cannot do the same with *the* horse or with a centaur. We can imagine the centaur, but we cannot touch it; *the* horse cannot be imagined, nor touched. Of what disposition, race, age, sex, can *the* horse be? Since it does not possess any of these characteristics, we cannot imagine it; we can, on the other hand, think of it. Because we are capable only of thinking of it, and not riding it we know that *the* horse is a concept, and not a real living animal.

The relationship, then, or the dealings we have with an object, reveal its nature to us. Well, then, what can we do with values?

Max Scheler maintains that intelligence is blind to values, i.e., it cannot have any sort of direct dealing with them. Values are revealed to us, according to the widespread theory of this German philosopher, through emotional intuition. Intuition is accurate and has no need to base itself on prior experience, nor on its corresponding vehicle. "We know of cases in the apprehension of values," he writes, "in which the value of a thing is given to us clearly and evidently, even *without* having the *carriers* of that value revealed to us."[1] José Ortega y Gasset, who made known Scheler's axiological conception to the Spanish-speaking world, wrote in 1923:

> The experience of values is independent of the experience of things. Moreover, it is of quite a different sort. Objects, realities, are by nature *opaque* to our perception. There is no way whereby we can ever see an apple in its entirety: we have to turn it, open it, divide it, and we shall never get to perceive it wholly. Our experience of it will continually improve, but it will never be perfect. On the other hand, the unreal—a number, a triangle, a concept, a value—is a *transparent* entity. We see them all at once in their entirety.[2]

Do we really perceive values at first sight and in their entirety? Are they really transparent? Are they revealed to us through emotional intuition?

The experience of artists, of art critics and historians, does not coincide with this optimistic description of the perception of value. A long and painful process is necessary at times for the work of art to disclose its previously hidden beauty. The grasp is never definitive; new approaches will afford us new surprises. In the realm of ethics, things are even more complicated. The honesty of one's behavior, or the injustice of a verdict, is not evident to us at first glance, and sometimes not even after long consideration.

We must be on our guard against the emotional character of the supposedly intuiting of value. Even within the esthetic realm, where the emotional aspect appears to predominate, there are intellectual elements which form part of our apprehension. If we proceed from the esthetic to the ethical or legal, the presence of rational elements is undeniable. When considering what is useful, reason takes the place of emotions. The utility of an object cannot be apprehended without a prior concept of the purpose which it is to fulfill, and the manner in which it fulfills it.

On the other hand, if it were true that we grasp values completely and intuitively, what should we do in the face of contradictory intuitions? There is not the slightest doubt that such intuitions exist. To say that one whose intuition is different from ours is blind to values implies arrogance and a lack of critical spirit; the clash of intuitions occurs in individuals of similar abilities. Which intuition will put an end to the intuitive contradiction?

These observations mean to point out the difficulties which every axiological theory should face up to, and the impossibility of eliminating the difficulties merely by affirming a point of view dogmatically. The problem remains

---

[1] Max Scheler, *Der Formalismus in der Ethik und die materiale Wertethik* (Bern: Francke-Verlag, 1954), p. 40.

[2] J. Ortega y Gasset, *Obras completas*, Vol. VI (Madrid: Revista de Occidente, 1947), p. 333.

wide open. The important thing is to understand the meaning, depth and complexity of the issue. Such understanding will prevent us from adopting a dogmatic doctrine, or from becoming disoriented in the face of contradictory theories which claim to be true and are supported by facts and reasons of similar weight.

THE DIMENSIONS OF VALUES

Nicholas Rescher

THE PROBLEM OF VALUE CLASSIFICATION

Why classify values? The question of the classification of values may strike the reader as a purely academic exercise of relatively little practical worth. But it is not so. One cannot begin a really coherent, well-informed discussion of any range of phenomena (dogs, games, diseases, etc.) until some at least rough classification is at hand. For classifications embody needed distinctions, and confusion is the price of a failure to heed needed distinctions. And, of course, in any practical application of the theoretical discussion—already at the level of gathering data—the mechanisms of classification are a virtually indispensable guide.

Owing to the inherent complexity of the concept of a value—the numerous and varied facets involved—value classification can be approached from many sides. Different principles of classification will, of course, be associated with these various angles of approach. This proliferation of perspectives from which values can be regarded serves in good measure to account for the absence of standard and widely accepted principles of value classification. Our own conception of value lays the basis for a diversified family of interlocking cross-classifications of values. It is fitting to set these out in some detail, both because it is useful to have these classifications at our disposal and because the consideration of such classifications sheds further light upon our understanding of the value concept.

It should perhaps go without saying that the handful of value classifications to be considered here is by no means exhaustive; these classifications are but key examples. The most that can be claimed for them is that they represent some of the most urgent requisites for a systematic survey of values that is to be precise in articulation on the one hand and fruitfully applicable on the other.

CLASSIFICATION BY THE SUBSCRIBERSHIP TO THE VALUE

Perhaps the most obvious classificatory distinction regarding values relates to the *subscribership* to the value. Is the value held—or is it such that it ought to

Nicholas Rescher, *Introduction to Value Theory*, © 1969, pp. 13-19. Reprinted by permission of Prentice-Hall, Inc., Englewood Cliffs, New Jersey.

be held—by a person or by a group, and then what sort of a group? Is the value (say, "self-esteem") put forward in the context of discussion as a value *of* Smith in particular, or *of* scientists in general, or *of* Paraguayans in general? Among what group of people is the value "at home," as it were; what is its appropriate setting? We correspondingly obtain such classificatory groupings as *personal* values, *professional* (professionwide) values or *work* values, *national* (nationwide) values, etc.

Like some of the other classifications to be introduced here, this is not a classification of values as such pertaining to their *content*, i.e., the subject matter with which the value deals. Rather, the value is taken as already given and fixed so far as its meaning-content is concerned; the only question is: Who holds it?

CLASSIFICATION BY THE OBJECTS AT ISSUE

In evaluation something is evaluated with reference to a certain valued characteristic: men, say, are evaluated *in point of* their intelligence, or nations *in point of* the justice of their legal arrangements, or precious stones *in point of* their purity. A value, say, "intelligence" or "justice" or "purity (in precious stones)," is thus correlated with a certain valued state of affairs in which it comes to be realized: men's being intelligent, society's being just, precious stones being pure. This state of affairs is specified in terms of the relevant mode of evaluation applied to the *object items* or *objects* at issue (men, nations, things).

Thus one of the ways of classifying values is with reference to the appropriate group of objects to which the value is taken to have application. Some of the main categories in a classificatory system of this type would be as follows:

Name of value type	Explanation of what is at issue	Sample values
1. Thing values	desirable features of inert things or of animals	purity (in precious stones) speed (in cars or horses)
2. Environmental values	desirable features of arrangements in the (nonhuman) sector of the environment	beauty (of landscape or urban design) novelty
3. Individual or personal values	desirable features of an individual person (character traits, abilities and talents, features of personality, habits, life patterns)	bravery intelligence (in man)
4. Group values	desirable features of the relationships between an individual and his group (in family, profession, etc.)	respect mutual trust
5. Societal values	desirable features of arrangements in the society	economic justice equality (before the law)

On the basis of this classification, it is readily seen that what is at first blush a single value may come to be differentiated with respect to its objects. So with "respect": *self-respect* belongs to category 3, but the *mutual respect* of associates falls into category 4. Again, the *justness* of an individual belongs to category 3, but the *systematic justice* of a nation falls into category 5.

The question at issue throughout this classification relates to the *domain of applicability* of the value. This is a matter of introducing—indeed of defining—the specific value at issue. Thus if what X values is "spaciousness in gardens," we cannot simply speak of "spaciousness" as one of the values he holds, for he might, for example, prefer compactness rather than spaciousness in, say, dwelling-houses. In indicating the domain at issue, then, we are actually *not so much classifying a given value as specifying which value is given*. These examples, moreover, show that the list given above is only a starting point. The specification of a domain lends itself to indefinitely greater ramifications.

CLASSIFICATION BY THE NATURE OF THE BENEFIT AT ISSUE

As a conception of the beneficial, a value is invariably bound up with a *benefit:* namely, that which is seen to ensue upon the realization of this value. Values can thus be classified according to the types of benefits at issue. To implement this prospect, we need to be able to effect a prior classification of benefits themselves. But how are benefits to be classified?

This question is a relatively simple one, for the notion of a benefit is correlative with that of human wants, needs, desiderata, and interests. Insofar as something conduces to the latter, it is to be classed under the rubric of the former. Thus, since we have a reasonably reliable view of human wants, needs, desiderata, and interests, we also have a plausible survey of potential benefits. This can be projected into a corresponding classification of values, somewhat along the following lines:

Category of value	Sample values
1. Material and physical	health, comfort, physical security
2. Economic	economic security, productiveness
3. Moral	honesty, fairness
4. Social	charitableness, courtesy
5. Political	freedom, justice
6. Aesthetic	beauty, symmetry
7. Religious (spiritual)	piety, clearness of conscience
8. Intellectual	intelligence, clarity
9. Professional	professional recognition and success
10. Sentimental	love, acceptance

In this classification we group values according to the generic *qualitative nature* of the benefit they involve—e.g., being in good standing with ourselves (7) or with our group (4) or with our professional colleagues (9); having welfare of mind (8) or of body (1); enjoying pleasantness in the condition of labor (2), or of life (8), or in the attractiveness of the environment (6). The guiding concept of this group of classifications is to differentiate values according to the nature of the benefit at issue—that is, according to the human wants, needs, and interests that are served by their realization.

We might, incidentally, compare the benefits at issue not in point of their *type*, but simply in point of their *magnitude*. This gives rise to another "dimension" of values, viz., their *height*. For one may assess values comparatively with respect to the extent of the benefits that accrue from their realization. It is just in this sense that "health" is a higher personal value than "comfort" or that the value of "courtesy" is lower in contrast to "justice."

CLASSIFICATION BY THE PURPOSES AT ISSUE

Values can also be classified with respect to the specific type of purpose served by realization of the valued state of affairs, as with *food value or medicinal value*, to cite paradigm examples. Thus we may speak of the *exchange value* of an artifact, i.e., its value for exchange purposes or again of the *deterrent value* of a weapon, i.e., its value for purposes of deterrence. Comparably, we might speak of the *bargaining value* of a certain resource, or even of the *persuasive value* of an argument. Again, monetary value can be grouped in this class: the monetary value of something being its value for purposes of acquiring money, its exchange value *vis-á-vis* money. (On the other hand, if *monetary value* is construed, as is perhaps more natural, as the value of something *measured in monetary terms*—so that the units of comparative evaluation are at issue— then a yet different mode of value classification is clearly at hand.)

Whenever the value under discussion is characterized in terms of the formula "value for _____ purposes," then we have a classification that proceeds with a view to describing the *mechanism* through which the benefit at issue in the value is to be realized (be it exchange, bargaining, persuasion, etc.). Again, something of the same sort is at issue with a locution of the type: "His counsel was of great professional value to me." For this way of speaking is presumably to be construed as: "His counsel was of great value to me for professional purposes, i.e., in facilitating my attainment of some of the elements of professional success." Here then, the classification of values proceeds in terms of groupings of the *loci* of value that are at issue, that is, the specific human purposes to the attainment of which the value is relevant. This mode of value classification would come to the fore in the analysis of a statement of the type: "The literary value of Samuel Pepys's *Diary* is limited, but its documentary and historical value is enormous."

CLASSIFICATION BY THE RELATIONSHIP BETWEEN THE SUBSCRIBER AND THE BENEFICIARY

In general, a person subscribes to a value because he sees its realization as beneficial to certain people. Consequently yet another approach to the classification of values takes its departure from this point and classifies values according to the "orientation" of the value, that is, according to the relationship that obtains between the person who holds the value, the subscriber, on the one hand, and on the other, the presumptive beneficiaries who benefit from the realization of the value.

This approach leads to a classification of the following sort:

I. Self-oriented (or *egocentric*) values
    (Examples: "success," "comfort," or "privacy"—that is, one's own success, comfort, or privacy)
II. Other-oriented (or *disinterested*) values
    A. Ingroup-oriented (or parochial) values
        1. Family-oriented values (family pride)
        2. Profession-oriented values (the good repute of the profession)
        3. Nation-oriented values (patriotism)
        4. Society-oriented values (social justice)
    B. Mankind-oriented values
    (Examples: aesthetic values or humanitarian values in general)

Again, this classification can serve in the identification (or individuation) of certain blanket values that spread over a wide area. For example, consider the fact that the rubric value "loyalty" covers such rather diversified values as *family* loyalty, or *professional* loyalty, or *national* loyalty.

CLASSIFICATION BY THE RELATIONSHIP THE VALUE ITSELF BEARS TO OTHERS

Certain values are viewed as systematically subordinate to others. For example, "frugality" can scarcely be viewed as a self-subsistent value, but as subordinate to "wealth," or to "self-sufficiency," or to "simplicity (of life style)," or some such other value or combination of values. Or again, "generosity" may be prized for its conduciveness to the "happiness" of others. In such cases, the benefit seen to reside in a realization of the value is looked upon as residing in some other, "larger" value or values to which the initial value is subordinate. Values of this subordinate, other-facilitating sort may be characterized as *instrumental* values or *means* values. Such values come to have this status not so much (or at any rate not primarily) because of their inherent nature, as because their realization conduces to the realization of other values that are regarded as more fundamental.

A second category of values stands in contrast to these instrumental values. The values of this second category are not viewed as subordinate but as self-sufficient. "Loyalty" or "honesty" are, in general, to be prized primarily on their own account, not because acting on them conduces to the realization of other, "larger" values. These, then, are the *intrinsic* values or *end* values. The benefit of realizing a value designated in this manner is seen to reside primarily in this realization of itself and for itself.

FURTHER CLASSIFICATIONS

The preceding list by no means exhausts the range of perspectives from whose standpoint values can be classified. For example, one could also classify values by *how soon the benefit at issue comes to be realized*. "Familiarity with the terrain was of great *immediate* value to him," "A knowledge of French will be of *ultimate* value to her," that is, of value to her in the long run. Here the value of the item being evaluated is appraised in terms of the question: How immediate or deferred is the acquisition of the benefits to be accrued from bringing the valued state of affairs to realization?

CONCLUSION

Six main principles for classifying values have been examined. We have seen that values can be differentiated by:

1. their subscribership
2. their object items
3. the sort of benefits at issue
4. the sort of purposes at issue
5. the relationship between subscriber and beneficiary
6. the relationship of the value to other values

These six factors indicate distinct "dimensions" with respect to which values can be characterized so that some of their key features can be set out in a systematic fashion. They provide a relatively clear and precise mechanism for discussing significantly general and persuasive aspects of values. By the use of such classifications at least part of the enormous complexity of values can be reduced to orderly terms. It is, moreover, useful to heed these different dimensions of value because an awareness of the distinctions that underlie them enables us to avoid invitations to confusion in value discussions. At first thought, it might seem that when one speaks of "societal value," "aesthetic value," and "exchange value," one is talking about the same sort of thing—just so many different ways of classing values with respect to a long but homogeneous list of distinguishing labels. That this impression is quite incorrect— that very different sorts of things are at issue here—has become clear in considering the highly variegated "dimensions" of value.

OBJECTIVES OF VALUE ANALYSIS

Jerrold R. Coombs

Having students analyse, discuss, and decide value questions, particularly those about which there is public controversy, has recently become the subject of renewed concern among social studies teachers. Despite their concern, many teachers cannot operate effectively in this area because of confusion and uncertainty. They are confused as to what, if any, legitimate educational objectives are to be obtained by such value analysis. This in turn produces uncertainty about procedures to be used by the teacher in directing value analyses, and about the means of evaluating student achievement resulting from such exercises. The purpose of this chapter is to make a start at clarifying the objectives of value analysis.

SOME PRELIMINARY CLARIFICATION

If we are to become clear about objectives, we must begin by being clear about the terms we use in talking about value analysis, particularly the term "value." Very often the term "value" is used in such a way as to be ambiguous. For example, in some contexts it may refer either to the things people hold to be of worth or to the standards by which people judge the worth of things. To avoid confusion we will use the term only in the phrase "value judgments." Value judgments may be defined roughly as those judgments which rate things with respect to their worth. The following statements express value judgments.

1. Nixon is a good president.
2. Washington is a beautiful city.
3. Capitalism is an efficient economic system.
4. War is mass murder.
5. The sinking of the *Titanic* was a disaster.
6. Proportional representation is an adequate way of giving voice to the will of the people.
7. The U.S. ought to stop testing nuclear weapons.
8. Presidents should be elected by direct popular vote.

From J. E. Coombs, Rational Strategies and Procedures. *National Council for the Social Studies Yearbook*, 1971, (41), 1-28.

Words such as "good," "beautiful," "efficient," "murder" are called "evaluative terms" or "rating terms" because they are commonly used to rate things with respect to their worth. The terms "ought" and "should" are also evaluative terms when they are used in prescriptive statements, i.e., statements telling us what to do. Such statements can be translated into statements containing more obvious evaluative terms. If some action ought to be taken or should be taken, then it is either right or desirable to take the action.

We will refer to the thing being rated in the value judgment as the "value object." Almost any sort of thing can be a value object. We evaluate physical objects, events, people, actions, institutions, and practices as well as classes of such things. In the statements above, Nixon, Washington, capitalism, war, sinking of the *Titanic*, proportional representation, testing nuclear weapons, and direct popular election of presidents are all value objects.

Value judgments may contain positive evaluations, negative evaluations, or neutral evaluations. For example, statements 1 and 2 express positive evaluations, statements 4, and 5 express negative evaluations, and statement 6 expresses a neutral evaluation. A positive evaluation places the value object high on some scale of worth, a negative evaluation places it low, and a neutral evaluation places it around the midpoint.

Evaluative terms vary with respect to how much they tell us about the value objects to which they are applied. Some evaluative terms such as "good" and "bad" tell us nothing definite about the characteristics of value objects, while other terms such as "murder" give us quite a bit of information. The judgment that Nixon is a good president tells us nothing definite about President Nixon. But the judgment that war is mass murder does tell us something about war, namely that it entails deliberate killing on a large scale.

One other feature of value judgments is worthy of mention. There are several different points of view according to which we assess value objects. We may assess a value object from an aesthetic, a moral, an economic, or a prudential point of view. Other points of view can be identified, but these are the most important.[1] In addition, we may make an overall judgment of the worth of the value object, taking into account various points of view. Some evaluative terms ordinarily are used to make assessments from only one point of view. For example, "beautiful" and "ugly" ordinarily rate things from an aesthetic point of view. "Efficient" rates things from a moral point of view, and "wise" conveys a prudential rating. Most evaluative terms can be used with reference to more than one point of view. "Good," "bad," "desirable," and "undesirable" can be used to rate things from virtually any point of view. We shall have more to say about the various points of view in a later section of this chapter. We will be particularly concerned with the moral point of view since it is the one which causes most confusion.

Value judgments are diverse and complex, having many guises and many functions. What has been said so far is not meant to be an exhaustive portrayal of such judgments. However, it should help us avoid confusion in discussing the objectives of value analysis.

RELEVANCE OF THE LOGIC OF JUSTIFICATION

Discussions of value analysis in an educational context suggest at least four possible objectives of such an enterprise.

1. To teach students that some value object is to be given a particular rating; to teach, for example, that the U.N. is a good thing or that premarital sex is wrong. This sort of objective is what people appear to have in mind when they advocate teaching values or teaching citizenship.
2. To help each student make the most rational, defensible value judgment he can make about the value object in question.
3. To equip students with the capacity and inclination to make rational, defensible value judgments.
4. To teach students how to operate as members of a group attempting to come to a common value judgment about some value object.

Teachers have to determine which, if any, of these proposed objectives are defensible educational goals. This implies that they also must know with some specificity just what they are attempting to accomplish in each case. What exactly is one attempting to accomplish when he sets out to teach students to make rational value judgments? When teachers are clearer about these things, they should be in a much better position to consider the kinds of procedures they should use in promoting and guiding value deliberations. Of course, having a clear understanding of one's goals is not a sufficient basis for determining what teaching procedures to use. It is necessary as well to have empirical knowledge indicating the kinds of procedures which are likely to produce the results we want. Chapter Two of this yearbook will attempt to assess the current state of the relevant empirical knowledge.

In claiming that teachers have to determine which goals of value analysis are legitimate, I do not mean to imply that teachers should be the final arbiters of educational goals. While I do not wish to argue the point here, my own view is that, in the final analysis, the authority for such decisions should rest with the community served by the educational institution. But, the community, aside from a few vocal minority groups, does not make clear in any detail what it expects of the schools. This means that, like it or not, teachers must make decisions about goals. However, they must have good grounds for their decisions, and they must be prepared to justify them to the community.

It is unlikely that teachers could come to well-founded decisions about the objectives of value analysis without some understanding of the nature of value judgments and the way in which they are justified. Consider the first objective on the list above, namely the objective of teaching particular value conclusions. If value judgments can be shown to be true or false in the same manner as judgments about matters of fact, a strong defense for adopting this goal is at hand. We can offer the same grounds for teaching a particular value conclusion as for teaching a factual conclusion, namely that it is true and important.

However, if evaluative judgments are, as some have claimed, merely expressions of emotions, preferences, attitudes, or tastes, we shall need a different sort of justification for teaching particular value conclusions. Indeed, we may find no justification for adopting this sort of objective for value analysis. Notice also that the third objective on the list, that of teaching students to make rational value judgments, is defensible only if the nature of value judgments is such that it makes sense to regard them as rational or irrational.

Understanding something of the logic of justifying value judgments is important as well in becoming clear about what exactly is implied by each of the proposed objectives. For example, the objective of teaching students to make rational value judgments implies that we will teach them to follow some standards of rationality in their judgments. Thus getting clear in detail about this objective is dependent in part on getting clear in detail about the standards of rationality applicable to value judgments.

Although clarifying the nature and justification of value judgments is an important first step in unravelling the confusion and uncertainty surrounding the objectives of value analysis, it is by no means an easy job. Many very good philosophers have been working on this task over a period of hundreds of years. Still there are things about value judgments which we do not know or cannot agree upon. What I shall have to say on this matter is neither new nor original. My justification for saying it here is that to my knowledge there is no single work which covers all the things that need to be said in a manner which is concise, intelligible to non-philosophers, and directly relevant to the problem of clarifying objectives of value deliberation. While these views are not entirely noncontroversial, they are supported by arguments which I consider compelling.* The reader is not asked to accept the conclusions on anyone's authority, but rather on the basis of the arguments offered in support of them. Because the arguments presented here are of necessity brief, a selected bibliography is provided at the end of this chapter for those who wish to delve more deeply into the issues.

Some people hold the view that any assertions about the nature of value judgments and how they are justified must themselves be value assertions expressing the author's basic values. Thus they contend that no account of justification is free of presuppositions themselves requiring justification. While there is a grain of truth in this view, in the main it is mistaken. It is possible, though difficult, to describe how we justify value decisions without thereby making any value assertions. This is what the present discussion will attempt. Briefly, what we want to uncover are the rules which govern our reasoning about value questions. Thus our inquiry is similar to that of the philosopher of science who attempts to map the rules governing our reasoning about empirical matters. However, as Paul Taylor points out, we are not concerned merely with describing how the majority of people make and justify value decisions. A great many people, perhaps most, do not always think rigorously in making value decisions. Consequently, we want to find out what

---

*Some issues regarding the nature of value judgments are still *very* controversial. No attempt is made here to resolve all such issues.

rules people follow when they are aware of how they are reasoning and are satisfied with it.[2] In a sense, then, we want to find out the rules which govern satisfactory or rational reasoning in the realm of making and justifying value judgments. "Rational" here means that manner of operating which one accepts when he has full knowledge of what he is doing. It is important to be clear that we are not *proposing* canons of rationality for reasoning about value matters. We are attempting to make explicit the ideal of rationality implicit in our use of value language.

VALUE JUDGMENTS AND FACTUAL JUDGMENTS

There are several widely held views about the nature of value judgments. Since it is likely that many teachers hold one or another of these views, it may be well to begin our discussion by considering these popular conceptions. One such conception holds that statements expressing value judgments are not significantly different from factual statements. The two types of statements may be shown to be true or false in the same way. This view has a certain amount of face validity. Value statements, like factual statements, appear to tell us something about the world of experience, and it seems sensible to assert that they are true or false.

The deficiencies of this view begin to show when we attempt to verify value statements by means of the procedures used in verifying factual statements. These verification procedures are of three types, designated here as VP1, VP2, and VP3.

VP1 is used only when the statement to be verified is an observation statement, i.e., when it reports a particular observable condition. The procedure consists in making an observation to find out if the condition reported in the statement is present. Using this procedure one might verify the statement "There is a pen on my desk" by looking or feeling on my desk to find out if there is a pen on it.

VP2 is the procedure of verifying a factual statement by deducing it from other true factual statements. If a statement can be shown to be the conclusion of a valid deductive argument having true premises, the statement must be true. For example, we might attempt to verify the statement "This vase is 2000 years old" by showing it to be the conclusion of an argument with the following premises. Artifacts with chemical property X are 2000 years old. This vase has chemical property X. Since this is a valid argument, the conclusion would have to be regarded as true if the premises were true. A second version of VP2 allows definitions as well as factual statements in the premises. In this version we might verify a statement such as "X is fragile" by deducing it from true factual statements about X plus a definition of the term "fragile."

VP3 is used to verify generalizations and involves two steps. First a statement of some observable condition is deduced from the generalization in conjunction with other known facts. Then observations are made to determine whether or not this predicted condition obtains The presence of the predicted condition provides evidence confirming the truth of the generalization. The absence of the predicted condition serves to falsify the generalization.

The question now to be settled is whether or not value statements can be justified in any of these three ways. Are value statements rating the worth of single objects, events, etc. capable of verification by means of VP1? Consider the value statement "That is a good car." Can this statement be verified by making observations of the car in question to determine if some condition reported in the statement is present? If we examine a car we observe color, size, shape, amount of gas used, stopping distance, etc. What we do not see is the goodness of the car. Nor do we taste, smell, hear, or feel it. We may *decide* on the basis of our observations that the car is good, but we do not observe goodness.

Were it the case that goodness could be observed, it would not be possible for two persons, both of whom had thoroughly tested and examined the car, to disagree as to whether or not the car is good. But in fact such a disagreement is not only possible, it is a common occurrence. Indeed it is a general feature of evaluative reasoning that persons can and do make contrary evaluations about objects and events even when they have made the same observations.

In sum, value assertions do not merely report or describe observable conditions, and they cannot be verified by means of VP1.

Putting aside consideration of VP2 temporarily, let us see whether or not VP3 can do the job of verifying general value statements. It can do the job only if it is possible to deduce from the evaluative statement in conjunction with true factual statements a statement describing observable phenomena. In the case of many general evaluative statements such a deduction is clearly not possible. Consider the value statement "Fighting is bad." This statement in conjunction with the factual statement "X is fighting" implies "X is bad." But "X is bad" does not describe observable phenomena. This case illustrates a significant point about a large class of value assertions. The only inference from this sort of general value assertion that is logically relevant to verifying the assertion is a value statement; a statement which does not describe observable phenomena. Therefore, many evaluative statements cannot be verified by means of VP3.*

Many people hold the view that value assertions can be verified by means of VP2. The attraction of this view stems from the fact that in some sense we do derive our value conclusions from factual considerations. In a previous example we noted that we cannot observe the goodness of a car, but it does seem that we conclude that a car is good on the basis of facts we know about the car. Indeed, though this is letting the cat out of the bag prematurely, the whole task of clarifying the nature of value judgments is to describe exactly how we reason from facts to value conclusions.

---

*Some singular value assertions can be shown to be *false* by VP1, and some general value assertions can be shown to be *false* by VP3. This applies only to those assertions containing rating terms such as "murder" which give information about characteristics of the value object. It does not apply to evaluations using primary rating terms such as "good," "bad," "right," "wrong," "ought," etc. *No* value statement can be shown to be true by VP1 and *no* general value statement can be significantly confirmed by VP3.

There are, you will remember, two versions of VP2: (1) deducing the statement to be verified from true factual statements and (2) deducing the statement to be verified from true factual statements and definitions. The first version can be used to verify statements only if it is possible to construct a *valid deductive* argument having only factual statements as premises and an evaluative statement as a conclusion. No such argument has yet been constructed.* The second version of VP2 can do the job of verifying value statements only if it is possible to define evaluative terms accurately by means of a definition itself containing no evaluative terms. No such definition has yet been produced.

There are good grounds for believing that the kinds of arguments and definitions which would allow us to use VP2 in verifying value statements are not possible. They are not possible because evaluative assertions serve functions not served by factual assertions. Evaluative assertions serve to tell us what to do; they provide guidance as to how persons are to act, choose, or feel about something. Factual assertions by themselves do not perform these functions. Thus definitions which permit the translation of value assertions into factual assertions cannot be accurate. And since the conclusion of a valid deductive argument asserts nothing that is not implied in the premises, arguments from factual assertions to evaluative conclusions cannot be valid.

VALUE JUDGMENTS AND ATTITUDES

A second widely held but inadequate view of value judgments may be characterized as follows. An evaluative statement is merely an expression or indication of the attitude or feeling of the person making the statement. If one says that medicare is a good thing, he is merely expressing his attitude of liking, approving, or being favorably disposed toward medicare. Further, it does not make sense to consider attitudes as being true or false, correct or incorrect, justified or unjustified. We can justify beliefs or show them to be correct by citing relevant facts which serve as evidence for the beliefs. We cannot give facts to show that attitudes or feelings are correct or justified. What facts, for instance, could show that it is correct to dislike blondes or that one is justified in enjoying football? Since evaluative statements are merely expressions of attitudes we only delude ourselves if we think they can be shown to be correct or justified. True, we do engage in arguments which purport to justify our evaluative assertions. But the only real purpose these arguments serve is to persuade others to adopt the same attitude we have.

This point of view has an element of truth in it, but it is faulty in several important respects. It is true that evaluative assertions imply that the person making the assertion has a certain attitude. If I say that school desegregation is a good thing, I am indicating that I have a positive or approving attitude toward school desegregation. It is also true that arguments given in support of

---

*Arguments of this type are often persuasive, but they are not valid unless some value assertion is added to the premises. They persuade us because we accept the suppressed (unstated) value premise.

my evaluation serve to persuade others to adopt my positive attitude or to persuade me to maintain my positive attitude. The deficiencies of this view lie in the conclusions that evaluative assertions are *merely* expressions of attitudes, and arguments supporting evaluations are *only* attempts to persuade people to adopt a given attitude.

It is easy to show that evaluative assertions are not merely expressions of attitudes. We do not make the same kind of investigation in deciding whether or not to accept or believe an evaluative assertion as we do in deciding whether or not to accept an expression of attitude.

Consider the following sentences.

1. U.S involvement in the Vietnamese war is wrong.

2. I disapprove of U.S. involvement in the Vietnamese war.

Sentence #1 is an evaluative statement; sentence #2 is an expression of attitude. To determine the acceptability of the first sentence we would gather information about the U.S. involvement in Vietnam and its consequences. The acceptability of the second statement is not decided in the same way. Rather, we would attempt to assure ourselves that the speaker was not deceiving us about his attitudes or feelings concerning Vietnam.

The view that arguments given to support evaluations are only attempts to persuade persons to adopt a given attitude has several weaknesses. First, it is wrong in assuming that attitudes cannot be justified. We can and do give reasons to justify attitudes. To justify disapproving of U.S. involvement in the Vietnamese war we might give as a reason the fact that it increases the rate of killing in the war. This reason is not meant merely to explain why we have the attitude but to show that the attitude is acceptable.

The second weakness in this view is that it does not allow us to make a very important distinction which we commonly do make in considering evaluative arguments, that is, the distinction between relevant justifying reasons and illegitimate attempts to persuade. In attempting to decide the worth of something we sort out persuasive considerations on the basis of whether they are legitimate and relevant or not. A teacher might want to give a student a good grade because the student is friendly and appealing, but he generally regards these considerations as irrelevant. This distinction between relevant and irrelevant considerations is learned in the act of learning to use value language. When a child states as his reason for hitting someone "I don't like him," we are apt to tell the child "That's no reason for hitting him." Thus we teach him that some considerations are not relevant, i.e., do not count as reasons. Any adequate account of evaluative reasoning must take into account this important distinction.

In sum, the view of evaluative judgments being considered here does not do justice to the fact that evaluative assertions are meant to express *justifiable* attitudes and they are meant to have *legitimate* authority over the conduct and attitudes of others. When we say that space exploration is a good thing, we not only imply that *we* approve of space exploration and that we want others to approve of it; we also imply that there are compelling reasons for *anyone* adopting a positive attitude toward space exploration.

FEATURES OF EVALUATIVE REASONING

It is time now to provide a more positive account of the nature and justification of value judgments. One of the most significant points to note is that it is always relevant to ask for justification of value judgments. That is, it is never beside the point to ask for reasons or grounds for the judgment. If someone asserts that racism is bad, it is relevant to ask why it is bad or what makes it bad. Two sorts of information can be given in response to this question. The evaluator may respond by giving facts about racism in light of which he regards it as bad. He may say, for instance, that racism is bad because it produces needless suffering. Alternatively, the evaluator may answer by citing a value principle such as "It is bad to have people suffer needlessly." A full specification of the reasons for any value judgment contains both facts about the thing being evaluated and rules or criteria which relate the facts to the rating. A person making an evaluation commits himself to having supporting facts because value judgments are logically dependent to a degree on factual considerations. One cannot apply a given rating to one value object and not to another if there is no difference in the facts about the two objects. One cannot, for example, say that table A is good and table B is not good if the two tables are exactly the same. Evaluations can differ only when there are differences in the facts. It must be kept in mind, however, that value judgments are not dependent upon nor derivable from facts alone.

Making a value judgment commits the evaluator to a value principle because his judgment logically implies a principle. If someone says that this pencil is good he commits himself to the value principle that any pencil just like this one is good. It would be logically inconsistent to assert the judgment and deny the value principle. The precise nature of the value principle implied by any judgment is indicated by the facts which are given to support the judgment. Suppose someone says this is a good pencil because it writes smoothly and feels comfortable in my hand. He commits himself to the principle that any pencil which writes smoothly and feels comfortable in my hand is a good pencil. If smoothness and comfort are what make this a good pencil, it follows that any pencil with these same features must be regarded as good.

The value principle implied in any judgment relates the supporting facts to the evaluative term used in making the judgment. In the example above the value principle relates facts about comfort and smoothness to the evaluative term "good."

To summarize, anyone making a value judgment commits himself to: (1) a value principle, and (2) a set of facts about the value object which shows that the principle applies to the value object. The facts and the value principle comprise the premises of a deductive argument having the value judgment as its conclusion.

While the rudiments of evaluative argument described above are rather simple, the actual process of evaluative decision making can be rather complex. We arrive at an evaluation on the basis of relevant facts. To be relevant to a value decision, facts have to meet two conditions: (1) they must be facts about the value object, and (2) they must be facts to which the evaluator ascribes some value rating. If the value decision is being made from a particular point of view, the value rating ascribed to the facts must be from the same point of view. The first condition needs no elaboration. The second condition may be equally obvious to some, but it is deserving of some explanation and illustration. Suppose someone were trying to decide whether or not it is good to build freeways into cities. One of the facts he knows about building freeways into cities is that they increase the total number of cars in the city. Unless the evaluator ascribes some value rating to having more or fewer cars in cities, i.e., thinks it is desirable or undesirable, this fact is not relevant to his evaluative decision.

To have ascribed value to some class of conditions is to have accepted or established a value *criterion*. We begin to accept or establish value criteria from a very early age, and by the time a person enters school he has a very extensive set of such criteria. Some widely held criteria include: it is wrong to cheat, lie, steal, kill, and hurt other people; it is good to keep promises, pay debts, and be healthy. Value criteria do not indicate the way in which a certain type of condition is to be rated in all circumstances. They indicate only how the condition is to be rated in the main or "other things being equal." For example, most of us accept the value criterion that it is wrong to lie. But there are many times when it is right to lie, as when lying will save someone's life or diminish his suffering. Thus while this value criterion does not hold in all cases, it does hold in most cases, or when there are no exceptional circumstances.

Value criteria not only make facts relevant, they give valence to facts. That is, they determine whether the facts support positive or negative evaluations. Suppose I am trying to decide whether euthanasia is good or bad. My value criterion that killing is wrong gives negative valence to the fact that euthanasia involves killing. Thus this fact supports a negative evaluation of euthanasia, i.e., a judgment that euthanasia is a bad thing. Whether the value decision is simple or complex depends upon the nature and extent of the relevant facts. If, in light of one's value criteria all the facts have positive valence, or they all have negative valence, the decision is fairly simple to make. The decision is complex and difficult when the relevant facts have conflicting valence, i.e., when some of the facts indicate the value object is good and some indicate it is not good. Then the evaluator must somehow weigh the facts and come to a decision. Most evaluative decisions which become the focus of concern in Social Studies classes are of this latter sort.

R. M. Hare has characterized the procedure used in arriving at justified value judgments as a procedure of conjecture and refutation.[3] After considering the relevant facts we make a tentative evaluation operating largely at the intuitive level. Then we test the evaluation, attempting to show that it is defective. If the evaluation withstands all our attempts to show that it is poor, we accept it. If it fails the tests, we start again with another judgment. In testing the value judgment we assess both the relevant facts and the value principle implied by the judgment. In a complex judgment this principle may be rather complex, adjudicating the conflicting claims of the various criteria relevant to the judgment.

An example will help make this clearer. Suppose an evaluator is trying to decide whether or not the U.S. ought to withdraw from the war in Vietnam. He accepts the following facts (f) and criteria (c).

(f) 1.The war in Vietnam is primarily a civil war.

(c) 1. One country ought not enter into the civil wars of other countries.

(f) 2. U.S. withdrawal will result in a substantially reduced rate of killing.

(c) 2. It is wrong to kill or to cause a large number of killings.

(f) 3. U.S. withdrawal would reduce the level of civil strife in the U.S.

(c) 3. A stable, peaceful society is a good thing.

(f) 4. U.S. withdrawal would free U.S. resources which could be used to cope with pressing social problems in the U.S.

(c) 4. It is desirable for a society to have the resources available to handle pressing social problems.

(f) 5. U.S. withdrawal would result in a repressive, communistic society in South Vietnam.

(c) 5. Illiberal societies are undesirable and immoral.

(f) 6. The U.S. has committed itself to defending South Vietnam against takeover by the communists.

(c) 6. A nation ought to honor its commitments.

(f) 7. U.S. withdrawal would be construed as a sign of weakness and lack of resolve.

(c) 7. A nation ought not let others think it is weak or irresolute.

The evaluator comes to the conclusion that the U.S. ought to withdraw from the Vietnamese war. He comes to this decision on the basis of the first four facts listed above, and in spite of the last three. His judgment implies a complex value principle to the effect that a nation ought not be involved in a civil war to save a country from a repressive government if that involvement increases the level of killing in the war and diverts the nation's attention from pressing social problems. It is this complex value principle that must be tested and found acceptable to the evaluator.

The value principle implied by a value judgment should not be confused with value criteria. Value criteria are *brought* to the context of value decision making. A value principle *emerges* as a *product* of that decision. It is only after a value decision has been made and the reasons for it given that we know what value principle is implied by the judgment. A value judgment may call into play a number of diverse and conflicting criteria, but only one value principle is implied in any value judgment.

The value object commonly has a number of features or aspects. These are described by factual statements about the value object. Each value criterion provides the basis for evaluating *one particular feature* of the value object, giving either positive or negative valence to that feature. Thus the value criteria we bring to a complex value judgment enable us to evaluate each feature of the value object *separately*. They do not provide the basis for evaluating the value object as a whole. In contrast, the value principle implied in our judgment does apply to the value object as a whole and so provides the basis for our decision about the object. It is a complex principle which adjudicates the claims of the various diverse and conflicting criteria. It reflects the relative strength of the various criteria in the context of this evaluation.

STANDARDS OF RATIONAL VALUE JUDGMENT

It is now possible to specify in general terms the conditions which a value judgment must meet to qualify as rational or defensible.

1. The purported facts supporting the judgment must be true or well confirmed.
2. The facts must be genuinely relevant, i.e., they must actually have valence for the person making the judgment.
3. Other things being equal, the greater the range of relevant facts taken into account in making the judgment, the more adequate the judgment is likely to be.
4. The value principle implied by the judgment must be acceptable to the person making the judgment.

The first standard follows from the fact that value judgments are to a degree based on factual considerations. If someone is wrong about the facts, his judgment based on the facts may also be wrong. Suppose someone judges that capital punishment is a good thing on the grounds that it deters serious crime. If as a matter of fact capital punishment does not act as a deterrent, the evaluator has made a poor judgment. If the factual mistake were pointed out to the evaluator he would have to reconsider and perhaps change his judgment.

The second standard follows from the fact that a value judgment is based in part on certain of the evaluator's attitudes toward facts about the value object. If an evaluator misrepresents his attitudes towards the facts, especially to himself, he is liable to make a judgment he could not accept or defend were he to become aware of the misrepresentation. Thus his judgment is not as rational as it could be.

The third standard is implied by the dependence of value judgments on facts. Suppose someone were to judge that building freeways into cities is a good thing because freeways move persons and goods into and out of the city faster and with fewer chances of accidents occurring. He does not take into account the facts that building freeways increases congestion and air pollution in the cities. If these additional facts were pointed out to the evaluator he very well might change his original judgment. Again, this indicates that his original judgment was not as rational as it might have been.

The fourth standard of rationality derives from the fact that one cannot accept a value judgment and reject the value principle implied by it without involving himself in a logical contradiction.

OBJECTIVES OF VALUE ANALYSIS

It is now possible to begin to answer some of the questions about objectives of value analysis raised at the beginning of this chapter. These questions, you will remember, had to do with the meaning and legitimacy of four possible objectives.

1. Teaching students to rate a value object in a particular way.
2. Helping students to make the most rational judgment they can make about the value object in question.
3. Teaching students to make rational value judgments.
4. Teaching students how to operate as members of a group attempting to come to a common value judgment about some value object.

A person adopting the first objective may attempt to teach either of two sorts of value conclusions. He may attempt to teach that some *single* object is to be rated a given way, e.g., the present draft law is unfair. Alternatively, he may attempt to teach a value criterion, i.e., that some *class* of value objects is to be rated in a given way, other things being equal. Given the validity of the argument thus far, it seems unlikely that teaching the first sort of value conclusion could be upheld as a defensible, educational objective. This follows from two features of value judgments. First, it is only the process of evaluative reasoning, not the conclusion of it, that can be judged adequate or inadequate. Second, the standards of rationality specify, among other things, the way in which the evaluator is to relate his *own* attitudes and preferences, embodied in value criteria, to facts in coming to an evaluative decision. Thus the teacher may conclude that X is good using evaluative reasoning that meets all the standards of rationality. Still the student's value criteria might be such that he could never come to the conclusion that X is good as the result of rational evaluative reasoning.

The legitimacy of teaching value criteria is not easily decided. Nothing in our discussion provides convincing arguments for not teaching any value criteria. Each criterion must be scrutinized independently. There are, I think, good grounds for teaching some value criteria and no good grounds for teaching others. Arguments for and against individual criteria are beyond the scope of this yearbook. Persons interested in such arguments should consult the book by Peters listed in the bibliography at the end of this chapter.

As indicated earlier, the second and third possible objectives of value analysis are viable only if there are standards of rationality applicable to value judgments. Our inquiry into the nature of evaluative reasoning indicates that such standards can be specified. No elaborate arguments need be advanced to show that these objectives are also worth achieving. Increasing the rationality of the conduct of students has long been accepted as an important objective of education in our society.

Nothing we have discovered about evaluative reasoning indicates that the objective of teaching students how to resolve value conflicts is insupportable. Since value judgments can not be proved true or false, and since they are not of

such a nature that agreement is always more important than the value conclusion agreed upon, we can teach no techniques that will ensure success in
resolving value conflicts. Still there are some things we can teach which will
increase the chances that students will be successful without compromising
the rationality of their judgments. Rational resolution of value conflict requires that the disputants identify the source of the conflict. Students can be
taught the possible sources of disagreement such that they can pinpoint where
the controversy arises in any particular case. Conflict can arise over (1) the
truth of some factual claim, (2) the relevance of a given fact, (3) the valence of a
given fact, (4) the interpretation of a particular value criterion, or (5) the
acceptability of the value principle implied in the judgment.

Returning to the Vietnamese war example cited earlier, we see that conflicting value judgments could result from disagreement over the truth of the
factual claim that U.S. withdrawal would result in other nations thinking the
U.S. is weak. There may be disagreement about the relevance of the fact that
this would be the first time the U.S. military has been unsuccessful in a war.
Some persons may have criteria relating this fact to some evaluation; others
may not have such criteria. There may be disagreement over the valence of the
fact that U.S. withdrawal is likely to result in a communistic regime in South
Vietnam. Some persons may have criteria rating this fact positively, while
others may rate it negatively. Disagreement may arise over the interpretation
of the value criterion that one nation ought not be allowed to impose its will on
another nation by force. Some persons may interpret "nation" in such a way as
to regard South Vietnam as a nation. Others may interpret "nation" in such a
way as to regard South Vietnam as part of one nation including both North and
South Vietnam. Finally, a dispute may arise because some persons accept the
value principle implied by the judgment that the U.S. should withdraw from
the Vietnamese war while others cannot accept the principle.

Exactly what we are trying to achieve when we aim for objectives two and
three is made clear by our discussion of evaluative reasoning. Achieving the
second objective means getting each student to make a value judgment using a
reasoning process which meets, insofar as possible, all the standards of rationality described above. The third objective entails teaching students to adhere
to the standards of rationality in making future value judgments. This objective is much more ambitious than the second and undoubtedly cannot be
achieved by a single value analysis. Students must be taught how to gather
facts, how to determine the relevancy of facts, how to assess the accuracy of
factual claims, and how to test the acceptability of the value principles implicit
in their judgments. In addition they must be taught the disposition to do these
things when they make important value judgments.

Some elaboration is necessary to make clear what is involved in testing
the acceptability of the principle implicit in one's judgment. Several different
tests of these principles are possible. One test involves making the principle
explicit, imagining other situations in which it would apply, and deciding if
one can accept its application in these other situations. An example should
make this clearer. Teacher Novice has tentatively decided that Frank, a boy

with a winning personality, is a good student. The facts he gives as reasons for his decision are that Frank always does his assignments neatly and on time, and averages 55 percent on his examinations. In order to test his implied value principle, Mr. Novice explicitly formulates it as follows: any student who does his assignments neatly and on time and averages at least 55 percent on his examinations is a good student. He then tests it by seeing if he can accept it when it is applied to other students. John, a quiet, colorless boy, meets all the conditions specified in the value principle, but Mr. Novice just cannot accept the judgment that he is a good student. This means the principle is unacceptable to Mr. Novice and he must reconsider his judgment.

A second way of determining the acceptability of a value principle is by trying to find reasons, i.e., relevant facts and more general value principles, which justify it. If one can construct such a justification he has good grounds for accepting the principle. Consider this example. The unrestricted use of D.D.T. to control pests is judged as inadvisable on the grounds that it needlessly endangers the health of people and animals. The principle implied here is that it is inadvisable to act in such a way as to endanger the health of persons and animals needlessly. The evaluator then determines the acceptability of this principle by finding out if there are relevant facts and more general principles to support it. He judges that it is acceptable because he accepts the following fact and general principle which support it.

FACT: *Anything which endangers the health of animals and persons increases the possibility that they will suffer.*

MORE GENERAL
VALUE PRINCIPLE: *It is inadvisable to increase needlessly the possibility that human beings and animals will suffer.*

These first two tests can be used to determine the acceptability of the principle implied in virtually any sort of value judgment. The two tests discussed below are applicable only to *moral* judgments about actions or practices. The third means of testing the value principle does not entail making the principle explicit. It is accomplished by the evaluator's exchanging roles with other persons affected by the judgment. If he can still accept the judgment when taking these other roles he can accept the principle implied by the judgment.[4] In the case of the previous example concerning U.S. withdrawal from the war in Vietnam, the evaluator might take the role of a South Vietnamese. If he were a South Vietnamese, could he still agree that the U.S. ought to withdraw from the war? If he cannot agree, then he cannot accept the principle implied in the judgment and so must reconsider his evaluation. If he can still accept the judgment he may further test the principle by adopting the role of some other interested party. The most crucial test is provided by exchanging roles with the person likely to be most adversely affected by the judgment.

The fourth test for determining the acceptability of a value principle is similar to the third in that it does not involve making the principle explicit. In this test the evaluator asks "What would be the consequences if everyone (in this sort of circumstance) were to act in the way the value judgment recommends?" If the consequences of everyone's acting in this way are unacceptable, then the principle implied in the judgment is unacceptable.[5] Suppose someone tentatively decides it is all right for him to burn his trash in the back yard. The circumstances are that he wants to use the money he would have to pay a trash collector for something else. By itself his trash burning would not make the air detrimental to plant or animal life. To test the principle implied in his judgment he asks, "What would happen if everyone in this circumstance were to burn trash in his back yard?" He decides that the consequences of this would be air pollution detrimental to animal and plant life. Since this consequence is unacceptable to him, the principle implied in his judgment is unacceptable.

SOME FURTHER CONSIDERATIONS

In the first section of this chapter we noted that value judgments can be made from a number of different points of view, e.g., aesthetic, moral, prudential points of view. We also noted that one can make an overall judgment of the advisability, desirability, or goodness of a value object. Different sets of facts are relevant to judgments from different points of view. For example, a judgment of the efficiency of the anti-poverty program would take into account somewhat different facts than those taken into account in judging the morality of the program.

Judgments from different points of view are often interrelated. That is to say, coming to a decision about a value object from one point of view often helps us come to a decision about the value object from a second point of view. Suppose someone is trying to judge whether a public health program is a good thing from the moral point of view. In the course of his deliberation he decides that the program is inefficient. This judgment may be an important factor in his moral judgment. He may decide the program is immoral because its inefficiency allows suffering that could be avoided. Judging a value object from different points of view may also be helpful in coming to a decision about the overall worth of the value object. For example, if we are trying to judge the worth of a certain car, it is helpful to decide whether or not it is safe, economical, beautiful, and dependable. Value judgments which are component parts of another value decision we call "subsidiary judgments." In the example above value judgments about the safety, beauty, and economy of the car are subsidiary to the judgment of the overall worth of the car. Each of these subsidiary judgments sums up a certain range of facts about the value object from a given value point of view. When one is faced with a large, complex body of facts relevant to his judgment he may choose to reduce the complexity by making a number of subsidiary judgments, each from a different point of view. He must then weigh the various subsidiary judgments and make his primary, overall value decision.

Judgments from the moral point of view are of particular interest because in a sense they take precedence over judgments from all other points of view. A judgment that the value object is *the* morally right thing to do or that it is morally bad overrides all other judgments of the value object. Only when the value object is neither morally required nor morally prohibited do judgments from other points of view determine the overall worth of the value object. Forcible sterilization may be judged an effective way to control population. However, if we also judge the practice to be immoral, then our overall judgment must be that it is not a good practice. If we judge smoking to be neither morally required nor morally wrong, our judgment of its worth will be decided by considerations relevant to other points of view, e.g., health. Moral judgments are distinguished from other sorts of value judgments by the fact that they are based on equal and impartial consideration of the interests of everyone concerned.

Up to this point we have written as though value objects are always judged in isolation from other things. Actually many value judgments are comparative. We try to decide if the value object is better or worse than something else or if it is the best of some set of alternatives. For example, we may try to decide if socialized medicine is better than private prepaid medical insurance or if urban renewal is the best way to improve housing in slum areas. When value judgments are comparative, relevant facts are those which allow us to compare the value object with its alternatives. Value criteria and the value principles implied by the judgment are also stated in comparative terms.

Current fashion in educational theorizing encourages us to conceptualize educational objectives as being either cognitive behaviors or affective behaviors, or a set of behaviors both cognitive and affective. Testing student achievement is then viewed as a process of observing to see if students exhibit the appropriate behaviors in appropriate circumstances.

It is instructive to consider the objective of helping students make rational judgments about the value object in question from this point of view. Is our objective to produce a cognitive outcome in the student or an affective outcome? Are we perhaps attempting to produce some outcomes of each type? These questions are difficult to answer. Arriving at a rational value judgment engages both the cognitive and affective aspects of a person's makeup. It involves the evaluator's cognitive structure because it requires knowledge of facts and the ability to test them. It involves the affective makeup of the evaluator because it is dependent on commitment to value criteria and principles, both of which embody feelings, attitudes, and preferences. When a person holds a rational evaluative conclusion there are some things he knows and some things he feels. The difficulty is that we cannot specify that the student must know specific things A, B and C and that he must have feeling D. What he must know is dependent to a degree on what he feels. Conversely, what he feels is determined by what he knows. We come back to the point emphasized earlier: the outcome sought is that students will have acted according to certain standards in making their decisions. It is difficult to see how this outcome could be described in terms of the cognitive-affective behaviors dichotomy. Assessment of student achievement with respect to this objective must be based primarily on performance during value analysis, not on behaviors exhibited after it.

SUMMARY

The primary focus of this chapter has been to answer the question, "What exactly are the legitimate objectives of value analysis in the classroom?" We have attempted to answer this question by examining the logic of value judgment and justification. In so doing we have argued the following points.

1. It is possible to describe our use of value language and the rules governing our reasoning about matters of value without thereby making any value judgments.

2. Value judgments are neither judgments of fact nor mere expressions of attitude.

3. Standards of rational value judgment can be specified but they apply to the process of value decision making not to the product of such a decision. These standards include:

    a. The purported facts supporting the judgment must be true or well confirmed.

    b. The facts must be genuinely relevant, i.e., they must actually have valence for the person making the judgment.

    c. Other things being equal, the greater the range of relevant facts taken into account in making the judgment, the more adequate the judgment is likely to be.

    d. The value principle implied by the judgment must be acceptable to the person making the judgment.

4. Since standards of rational value judgment can be specified the following objectives of value analysis in the classroom are defensible.

    a. Helping students make the most rational, defensible value judgments they can make.

    b. Helping students acquire the capabilities necessary to make rational value decisions and the disposition to do so.

5. There are no logical grounds for deciding that value criteria ought never to be taught nor for deciding that resolution of conflict about value matters is an illegitimate objective of value analysis.

NOTES

[1]For an extensive discussion of points of view see Georg Henrik von Wright, *The Varieties of Goodness* (London: Routledge & Kegan Paul, 1963).

[2]Paul W. Taylor, *Normative Discourse* (Englewood Cliffs, N.J.: Prentice-Hall, Inc., 1961), pp. 115-117.

[3]R. M. Hare, *Freedom and Reason* (New York: Oxford University Press, 1965), pp. 87-92.

[4]This principle is suggested and discussed at length in *ibid.*, pp. 90-95.

[5]This principle is expounded in Marcus Singer, *Generalization in Ethics* (London: Eyre & Spottiswoode, 1963), pp. 61-96.

SELECTED READINGS
IN THE LOGIC OF VALUE JUDGMENT

Baier, Kurt. *The Moral Point of View*. Ithaca, New York: Cornell University Press, 1958.

Frankena, William. *Ethics*. Englewood Cliffs, N.J.: Prentice-Hall, Inc., 1963.

Hare, R. M. *Freedom and Reason*. New York: Oxford University Press, 1963.

Peters, Richard S. *Ethics and Education*. London: George Allen & Unwin Ltd., 1966.

Singer, Marcus G. *Generalization in Ethics*. London: Eyre & Spottiswoode, 1963.

Taylor, Paul W. *Normative Discourse*. Englewood Cliffs, N.J.: Prentice-Hall, Inc., 1961.

von Wright, Georg H. *The Varieties of Goodness*. London: Routledge & Kegan Paul, 1963.

Wilson, John. *Reason and Morals*. Cambridge University Press, 1961.

Wilson, John, Norman Williams, and Barry Sugarman. *Introduction to Moral Education*. Baltimore: Penguin Books, 1967.

VALUE AND FACT

Wolfgang Köhler

The problem which is implied in the title has two sides, because we use the term "value" in two different meanings. From R. B. Perry we have learned that when we deal with value we must clearly distinguish between two levels of discourse. To start with the lower level: In a general sense the death of a competitor will be a value to *some* persons, and so will revenge or undeserved praise. Also, some individuals tend to like blond individuals of the other sex, some find the taste of buttermilk very pleasant, and some are repelled by the smell of tobacco. So long as we remain on this first level of value, there is no reason why we should not continue the list with instances like these: I must keep my promise; it is bad to deceive one's friend in order to gain some personal advantage. Value in a generic sense, i.e., value mainly defined by its contrast with neutral facts, is an ingredient which the latter examples have in common with the former. But many philosophers feel that, nevertheless, two issues must here be sharply separated. On the lower level of value theory, they would say, there is no reference to validity or to obligation. Value experience itself, however, forces us to rise to a higher level, and to segregate from values in general a more specific group whose values claim to be binding. Once we are on this level, a secondary evaluation of values in general becomes necessary in terms of binding value. Many values are positive on both levels. On the higher level it remains, for instance, a good thing if we like nice people. But some values which have prima facie a positive sign, assume a negative sign from the higher point of view. It seems to be generally acknowledged that revenge is in this class. Again, other values appear now as neutral. Gentlemen are under no obligation to prefer blondes; on the other hand, they cannot be blamed if they do so.

In the following, I propose to concentrate on value in the general sense of the word. The problem of valid value will only be discussed so far as seems necessary to prevent a fairly common misunderstanding. There are two rea-

From *The Journal of Philosophy,* 1944, 41 (8), 197-212. Reprinted by permission of the publisher.

The author wishes to remark that the present version of this paper differs at several points from the original which he read on November 28, 1943, for the Symposium on "Value in a World of Fact," at the Conference on Methods in Philosophy and the Sciences, New School for Social Research. In this corrected version free use has been made of the criticism which members of the audience, particularly Professor E. Nagel and Professor C. G. Hempel, raised in the discussion.

sons for the restriction. In the first place, it has become customary to ignore certain difficulties which concern value in its modest general sense. In the second place, certain errors which disturb our thought on the higher level are not peculiar to this level but only persist because they are first committed on the lower level. Both reasons make it advisable for one to refrain from an immediate discussion of valid value.

According to the scientists, no object and no event in nature has value characteristics. It does not matter whether we refer to value in general or to binding value; the whole genus is excluded from the description of nature. Physical objects merely exist and physical events merely occur; such is the verdict of natural science. On the other hand, every bit of human conduct, from trifles up to the present war, involves value as its most important content. This constitutes our problem.

A few hundred years ago there was no such problem. Everybody knows that in Aristotelian physics value concepts were freely applied to nature, and that remnants of this attitude are clearly discernible in the statements of great scientists as late as the seventeenth century. It is nevertheless true that at the present time a physicist would risk his reputation if he were to use value concepts in relation to his subject matter. At the same time the dualism to which I just referred is nowhere more striking than in the case of the scientist himself. He will scorn as ridiculous any suggestion that value might reside in physical facts. But his conduct as a scientist implies firm belief in a whole set of values. Quite apart from practical applications, knowledge and research are to him matters of tremendous value. The objectivity of science, clearness of reasoning, and honesty in the treatment of observational data are all values about which he has stern convictions. Paradoxically enough, even the absence of value from the subject matter of science impresses him as a great value which he will defend at all costs.

We would not have to be particularly disturbed by this attitude if it had no further consequences. But during the past three hundred years science has, for good reasons, gained a prestige with which few other human enterprises can compete. As a result large sections of civilized mankind have learned to look at things with the eyes of science. They have also adopted the values of science, for instance, the conviction that a keen mind with strong theoretical interests is best applied if it is applied to science. Thus the most intelligent men of generations went into science, and consequently forgot about the problems of value. They *had* value; but apart from its absence in nature they were not concerned with the topic. Under the same influence philosophers, who actually kept outside, did far more penetrating work about our access to the outside world and related problems than they did about questions of value. Is it astonishing that in a period which had this orientation the study of value degenerated? After all, value was not merely ignored in science. Rather, complete absence of value from any scientific material was now a fact of which all scientists were firmly convinced. Therefore, when new disciplines developed which proudly adopted the ways of science, not only scientific techniques in a positive sense spread far and wide, but also the scientists' dislike of value as a component of any subject matter. This happened in biology, in psychology, but to a degree also in social science. Thus during the nineteenth century value tended to become a matter of contempt, i.e., a negative value, among the intellectuals. I remember distinctly that as a young student of science I once expressed my condescending sympathy with an historian of art because in his field the poor fellow had to put up with value. It is obvious that this sophomoric behavior merely reflected a view which was then widely held. No less a man than Professor Titchener excluded value from serious consideration in scientific psychology. According to him this science deals with strictly neutral facts, just as does physics. It is quite possible that future generations will look back upon this period in utter perplexity.

One feels inclined to assume that in the minds of some scientists the situation caused uneasiness. If it did, such misgivings were soon dissipated by a remarkable doctrine which claimed that value could be explained on the basis of assumptions which never left the field of science. This achievement appeared possible in the framework of Darwin's theory of evolution. According to Darwin, characteristics of organic life which look as though they implied value can nevertheless be accounted for in terms of neutral facts. Presently an interpretation of values was proposed, and widely accepted, which applied the form of Darwin's reasoning to this further subject.

In Darwin's theory a species develops and improves its organs by accidental variations of germ cells, and by the early elimination of individuals who stem from cells with less fortunate variations. The elimination is brought about by the environment, in which the less fit individuals die earlier. An extension of this principle to the case of value will take more or less this form: The equipment of man as to mental tendencies is just as much a function of the germ cells as his anatomical characteristics are, and his equipment in this respect will vary when certain chance variations occur in those cells. Now, man is a social animal, and he will survive longer if he has mental tendencies which make him fit for life in a group. Thus in the remote past those individuals had the best chances who came from cells in which such tendencies were strongly preformed. The others died earlier and had fewer offspring. Nowadays, therefore, most people come from the better equipped stock. But what do we mean if we say that a man's mental tendencies favor his living and surviving in the group? Obviously we mean what are otherwise known as his moral attitudes with regard to others. They preserve the group, and through the group, the individual himself.

We need not investigate whether the authors of the evolutionary theory of value believed that in this fashion value was actually deduced from categories of natural science. There is little doubt that the theory was generally understood in this sense, and that it still owes its great popularity among scientists to that claim. In actual fact, of course, the claim represents a remarkable instance of self-deception. Plainly, Darwin's theory derives biological phenomena from neutral facts only inasmuch as the conditions which accidentally vary in germ cells are, and refer to, such neutral facts. It may be that Darwin's strictly biological reasoning fulfills this prerequisite. But, surely, the same prerequisite is not fulfilled in the evolutionary theory of value. For if here dispositions for moral attitudes in group life emerge in germ cells, such curious newcomers have certainly no ancestry among the neutral facts which make up nature according to natural science. Dispositions of this kind may vary from one germ cell to another; but, if so, it is dispositions *for value* which vary; instead of being deduced from neutral facts, value per se is tacitly taken for granted and, without an explanation, simply added to such facts. Or does natural science give us any hint of how a neutral fact can become a disposition for value? In other words, merely the form of Darwin's reasoning is here preserved. This form causes the erroneous impression that the theoretical achievement is also the same, namely, the reduction of a puzzling phenomenon to principles of natural science.

This ought to have been clear almost a priori. In the scientific thesis that no value attribute applies to any facts of nature, it is clearly implied that neutral facts and values are two sharply distinguishable genera. If this is right, how can any sequence of neutral facts, be it ever so long and complicated, have value as its end product? A theory which holds that this happens in evolution *must* somewhere commit an error. To be sure, evolutionary theorizing may be able to explain that human conduct shows certain regularities, i.e., that under given conditions one activity rather than another is likely to follow. But as soon as human action implies "good," "bad," or "ought," we are dealing not with additional regularities of mere occurrences but with a new class of facts. Moreover, far from being influenced by philosophical prejudice, we make this sharp distinction because it is made so sharply in science. In fact, therefore, the scientists themselves ought to have been the first to object to the confusion of neutral fact and value which characterizes the evolutionary theory. Actually, the criticism which the theory so much deserves has come from philosophy. During the past forty years there has been a renaissance in the field of ethics, and more than one philosopher has made it clear that, wherever the error lies, if the evolutionary theory is accepted, value is necessarily denaturalized. Evolution conceived as a sequence of changes in the hereditary make up of our ancestors will never explain value. I wish to add that this statement holds quite generally. Although both the evolutionary theory and its critics have been principally interested in moral value, our argument against the theory holds for value in any sense, including its most modest specimen. And we may generalize still further. Quite apart from evolutionary theory, so long as the present strictly negative view of natural science with regard to value is maintained, no datum concerning the physical world, among others no fact of neurology or nerve physiology, can have any bearing upon the nature and the problems of value as such.

In the meantime it has proved impossible to exclude value from the investigations of at least one empirical discipline, namely, psychology. Many psychologists may never use the term. Value is none the less implicitly accepted inasmuch as, after the gradual collapse of pure connectionism and conditionism, *motivation* has been generally recognized as the central issue of psychological inquiry. Few will deny that in human subjects the goal of motivated behavior has value characteristics. Now, in our present argument this acceptance of value as an empirical subject appears little short of baffling. For if natural science acts as though the attribution of value to anything in nature would be almost a crime against the spirit of factual research, how can one form of empirical investigation ignore this view, and yet proceed with a considerable measure of success? It is quite true that psychology has long followed the orders of natural science; on the other hand, at the present time psychologists may not be fully aware of the fact that with their study of motivation they are out of bounds so far as natural science is concerned. But it seems surprising that the transgression has no evil consequences. Neither law and order on the objective side nor exact procedure on the side of method appear endangered when motivation, and with it value, are freely included in empirical research.

To be sure, psychology has not yet shaken off the old fetters entirely. For the most part psychologists hold a view of value which is at odds with value experience, and must be caused by a prejudice. The prejudice, it appears, still has its roots in the unwillingness of science to admit value as an ingredient of objective situations. The now prevailing theories of motivation are *subjectivistic*, which implies that the interpretation of value in psychology is equally subjectivistic. It is widely taken for granted that all motivation originates in the self, and, as a result, value is generally said to be valuation, i.e., to spring from the self's subjective attitudes. Perry, for instance, whose theory is primarily a psychological theory makes great efforts to prove that value is wrongly localized when it appears within objects.

I have often asked philosophers and psychologists who defend this subjectivistic view just why they are so insistent about this point. So far I have never received a satisfactory answer. Least of all do answers point to direct observations of value experiences. Behind the answers there always seems to lie some deep conviction which is never openly formulated. I am afraid the objective parts of human experience, more particularly, our percepts, are for most of us so closely related to the world of natural science that the verdict against value in nature is inadvertently extended to the percepts. Even in their theoretical thought, few people can keep percepts and physical objects sharply separate. Consequently, since physics is regarded as a realm of purely neutral facts, percepts, which in a numerical sense tend to be identified with physical entities, can also never have value attributes. If percepts are not actually identified with physical objects, they are at least interpreted as almost literal translations of patterns of physical stimuli into patterns of corresponding sensations. Nerve physiology seems to give an adequate account of the genesis of such sensations in terms of natural science, and thus it once more follows that in percepts there is no place for value. Apart from these and closely related arguments, I can find no reason for the persistence with which the subjectivistic interpretation of value is still being defended. That part of our experience which we call the self—with its moods, efforts, emotions, and so forth—is not so closely related to nature as investigated by physicists and chemists. Therefore it does not seem to matter if this entity is made responsible for value as merely one strange function among many others.

A time will probably come in psychology and philosophy in which it will be a principle of method that the obvious characteristics of primary observational data are to be respected at their face value, whatever their relation to general preconceptions may be. It will then be acknowledged that we are not allowed to interpret black as really white, and here as actually elsewhere, unless we are forced to do so by further convincing observations. The time in which this rule will be generally accepted is likely to be remote. But it can be brought nearer only by one attempt after another to do precisely what the rule demands. The rule asks for phenomenology.[1]

How, then, does value appear in common experience?

1. It seems to be a fairly general fact that value appears as an attribute of things and events themselves rather than as an activity of the self or as the result of such an activity. We should therefore falsify our primary observational data if we were to say that the essence of value is valuation. Phenomenologically, value is located in objects and occurrences; it is not an action to which they are subjected. Value may reside in the most varied classes of things. A dress may look elegant or sloppy, a face hard or weak, a street cheerful or dismal, and in a tune there may be morose unrest or quiet power. I admit, one's own self is among the entities in which values may reside. Such is the case when we feel fit or, at another time, moody. But the thesis that it is always valuation as an act which imbues its object with value as a pseudoattribute is perhaps nowhere more artificial than precisely in this instance. Here the self would have to equip itself with value attributes such as fitness or utter fatigue. The idea seems slightly fantastic. And if in this instance a thing per se manages to have value characteristics, why should we deny this possiblity where other percepts are concerned?[2]

To repeat: For what reason are we to accept the thesis that value in objects can only be understood as an illusion, and that the truth of the matter consists in acts of the self which we wrongly objectify? There are instances in which the thesis would ask us to believe what is clearly impossible. Men sometimes succumb to what they would call the irresistible womanly charm of certain persons. Now, this is a value attribute on which women have a monopoly. It would be absurd to maintain that when the intensely male interests of such men impinge upon the actually neutral appearance of the women, female

---

[1]To avoid misunderstanding it will be necessary to add that phenomenology in the present sense differs from Husserl's technique. I do not believe that we are justified in putting certain phases of experience in brackets. A first account of experience ought to be given and carefully studied without selections of any kind. It is otherwise to be expected that even if the brackets are introduced as mere methodological tools, they will sooner or later turn out to be weapons of an ontological prejudice. In fact, I am not sure whether Husserl himself has not used them as such weapons.

[2]It will not be necessary to explain that the self, taken in the simple sense in which fitness or fatigue may temporarily become its attributes, *is* a particular percept; in othe words, that it is neither the physical organism nor the epistemological ego of certain philosophies.

charm develops in these objects as an illusory projection of the males' conations. Female charm is not a component of such male conations. How, then, can it be projected? It has been said that in pure description value characteristics do reside in objects; that to this extent the objective location of those characteristics is not an illusion; but that they are nevertheless *caused* by conations of the self. This theory still rests on the tacit premise that it is more natural for value to spring from the self than to reside in things. And we can only ask again: Whence that premise? Moreover, even in this form, the act theory of value negates phenomenological evidence, particularly in the case of certain negative values. When suddenly an object looms dangerous and threatening, it is not felt to have that appearance because there is first fear and a tendency to escape in the self. On the contrary, fear with that tendency is felt to be a consequence of the danger that threatens from the object. The theory is therefore at variance with experience.

It is merely a construction even in those instances in which value within the object actually varies with the intensity of the corresponding subjective need. It may well be that female charm is greatly enhanced when the sex need in a male has grown particularly strong. Does it follow that the sexual conation of the male self is responsible for that intensified perceptual value? If we say so we ignore another possibility. Glandular action (which is physiological action and not sexual conation as a psychological attitude) may have two parallel effects upon the nervous system. Those parts of the tissue whose states are correlated with phenomenal states of the self may become ready for processes that go with sexual conation. But in other parts which contain the correlates of perceptual objects, such processes may at the same time be intensified as are the somatic counterparts of certain tertiary qualities. From this point of view the latter change would not be established *by* the conation but *together with* increased readiness for such conation. If it is asked why, say, in the case of sex that glandular action should emphasize just those *particular* characteristics of certain percepts, the obvious answer is the other question: Why does the same physiological cause make the self ready for sexual conation rather than for any other conation? The former connection is not a bit more problematic than is the latter.[3]

It might be argued that a theory of value should not be concerned with mere percepts and their characteristics, since ultimately value must be referred to objective reality. But this argument seems to me mistaken. In the first place, if objective reality is meant as a synonym of the physical world, the scientific view which is at present generally held excludes any such consideration. According to this view there can be no value in objective reality. In the second place, we do not actually refer our moral obligations (or values in general) to the physical world. For all practical purposes we are naïve realists, and percepts are our real objects. It is quite true that in referring to the values of such objects we often go beyond perceptual fact. For everybody's percepts are equipped with characteristics which are matters of indirect knowledge rather

---

[3]If for a moment we have here left the phenomenological realm, we have done so because the view under discussion is itself a hypothetical construction.

than of sensory experience. This "thought content" of percepts, however, just like the content of other knowledge, is phenomenally objective. It is located within the percepts, and if it has value characteristics these, too, have an objective location. Incidentally, when images have value characteristics, their value is also phenomenally objective. Images are not part of the self.

In concluding, I should like to suggest that further interpretations of value in terms of valuating acts be accompanied by a statement of the premises which make the author prefer that interpretation to the testimony of value experience. It may be that once these premises become clear, the preference itself will be greatly weakened.

2. So far our phenomenology of value agrees with the descriptive accounts given by several philosophers in England although, on the whole, their accounts refer to specifically moral and valid value rather than to value in the generic sense. But if the same authors regard value *merely* as a quality which resides in objects, they seem to me to ignore an important phase of their subject. To the extent to which value characteristics are located in things and events themselves, and need for their existence no activity of the self, a relational interpretation of value would be misleading. But if the independence of value is admitted to this degree, we have not yet touched upon another side of the value experience which we cannot omit without unjustified abstraction. Even though value does not depend upon acts of the self, it is often clearly referred to certain facts within the object which exhibits the value. Moreover, these facts are for the most part *relational* facts. Anybody who perceives a graceful object or gesture will on reflection admit that this is obviously true. Actually, the observation holds for most instances of value: a situation or an object is felt to be valuable with reference to certain structural characteristics of the situation or the object. In this sense Professor Ross appears to be right not only when he regards value as a "consequential" quality but also when he accepts Professor Urban's thesis that as a rule value adheres to a *Sachverhalt*. In this German term it is implied that the factual datum to which the value characteristic of a situation refers tends to be a structural or relational trait of this situation.

I find it difficult to decide whether the preceding statements concern merely frequent or actually necessary characteristics of value. But with regard to a further trait I have no such doubts. Professor Urban holds that value is not adequately described unless we mention what may be called a *demand* character that belongs to its very nature. The plus and the minus signs which are characteristic of all value do not solely indicate that value qualities as such lie in one direction or the opposite with regard to a neutral point. Rather the plus and minus also mean "to be accepted, reached, maintained, supported" and "to be avoided, eliminated, changed in the positive direction." I am not acquainted with values of which this is not true. No value attribute seems to deserve its name if it has no such demand character. Moreover, while natural science would hardly worry about any other qualities with which phenomenology chooses to deal, it is precisely this demand character of value, in a wide sense the "ought" or the "requiredness" in it, which science wants to exclude when it refuses to use any value terms in reference to its subject matter. It is in contrast with any "ought" that facts of science are called *neutral* facts.

While, therefore, value cannot simply be interpreted in relational terms, it nevertheless contains a reference beyond its existence as a quality and beyond the object in which it resides. This reference is dynamic. Now the act theory which we discussed in earlier paragraphs emphasizes such a dynamic ingredient in value. But while this theory tends to identify value with a "vector" which issues from the *self*, in actual experience it is first of all value in an *object* which goes with a demand. In other words, the vector issues from the object qua valuable. The dangerous object threatens, the cool drink is tempting, the problematic situation invites closer inspection, and so forth. Here again, therefore, we have to reverse the account which the act theory gives of value. It shall not be denied that the demand which issues from a value object is mostly, though perhaps not necessarily, directed toward the self, and that as a consequence the self is likely to become active toward, or away from, that object. In this case, it may rightly be said, vectors actually arise of which the self is the center. But as a rule this happens because, in the first place, there is a positive or negative demand in the object.

3. We have just used the expression "Value in an object goes with a demand." This expression is not quite adequate. Between the value as such and the demand there is more than a factual connection, as though the demand merely accompanied the value characteristic. For, generally the attraction (or the negative vector) which issues from the object is felt to spring from the very nature of that value attribute. So far as I know it was W. Dilthey who first introduced the concept of understandable relationships. The eighteenth century was very fond of the term "reason." Reason was then considered a mental faculty. At the present time we prefer to speak of rational or understandable relationships which are experienced within actual mental situations. By this we mean that a given content may be felt to be required by others. But there is one further point at which the rationalism of that period must be corrected. We often apprehend that one thing follows from another as its natural consequence not only in the strictly intellectual field but also in motivation, emotion, and value experience. An understandable relationship in thought may not in every respect be comparable with such a relation as experienced in those other instances. But, surely, the relation between anger and its object, or joy and the situation from which it springs, is also understandable in the sense that, as a rule, such emotions appear adequate with reference to the facts from which they derive. We should, for instance, feel it to be both against the nature of the emotions and their objects if the object of my joy were all at once, without a change of its character, to cause the emotion of anger. A relationship of precisely this type is meant if we say that demands which issue from value objects are felt to spring from the nature of those objects and their values. In other words, we understand it as sensible that we are attracted by objects which have certain characteristics, and that we are disgusted by others. These are not mere sequences; at least prima facie it is felt to be adequate that just such vectors issue from such objects. The limitation which is contained in the

term "prima facie" refers, of course, to the possibility that on the level of valid value it may become an obligation to resist such perfectly understandable demands. If this happens, the relation between the value objects and their demands may nevertheless remain understandable in the present sense.

It is at this point that a remark about the problem of valid value obtrudes itself. Plainly, this problem is whether the demand which issues from a situation may be experienced not only as understandable but also as binding. If there are instances of this kind, two further questions follow immediately: First, under what specific circumstances do understandable demands appear as binding? And, secondly, what in this connection is the meaning of the term "binding"? It is a serious blunder when in relativistic arguments such phenomenological questions are replaced by questions of inductive generality, as though any problem of validity could be solved by counting how many people or tribes agree, or disagree, with certain judgments. Validity of a demand is quite compatible with factual dissension concerning the same demand. We see this every day in the purely cognitive field; but in this field of thought we draw no relativistic conclusions concerning the validity of logical demands. Even if it is to be granted that validity in thought is not throughout equivalent to validity, say, in ethics, the comparison shows that as a form of reasoning the relativistic arguments confuse two entirely different issues.

After this exploration of generic value as such, we can now return to our main problem, the dualism of value and the neutral facts of natural science. So far it has been our impression that no bridge can span the gap between these two concepts. If this were actually true, value in any sense of the word would have jumped into our world from nowhere when the development of living organisms had reached a certain stage, namely, when percepts had for the first time inviting or forbidding value characters. In a period which is as eagerly concerned with unity of knowledge as ours is, this must appear as a disturbing thought. Nor does it help us if we are told that, after all, values are also facts, and that therefore we need not bother about the relation of facts and values. This is not a question of mere terms. If values are facts, they are none the less facts of a peculiar kind. Otherwise why should natural science exclude this kind from its domain? Thus, if the scientists are right, the dualism as such remains in full force whether or not we subsume values under the category of facts.

We should, perhaps, be less disturbed by the dualism if value were only a quality of a particular kind. Quite apart from value attributes, a great many qualities of human experience are strangers in the world of natural science. Everybody seems to be resigned to the fact that, even if these qualities may once gain a better place in the scheme of science than they now have, present thought appears utterly unable to imagine what that place could be. In this respect value characteristics qua qualities are hardly more enigmatic than sensory qualities such as green or gray. We will therefore not attempt to relate our first characteristic of value to the world of science. It should, however, be remarked that if value characteristics were interpreted as projections of valuating acts, the qualitative differences among these acts would be just as puzzling to natural science as value qualities which reside in objects and events themselves.

The second and the third points of our description concern natural science more directly. For these points refer to structural characteristics of value rather than to mere qualities. If a demand issues from a thing and affects another, this is a structural fact; and if this demand is experienced as springing from the nature of its substratum, this is once more a matter of functional structure. Here science finds itself in a curious position. While it excludes value from nature, it postulates at the same time that the structural characteristics of all functions in the world be interpretable in terms of physical facts. This will become clear if we once more return to the principle of evolution. Evolution has two sides. It is commonly regarded as a principle of *change*, because evolution has established one new anatomical condition after another. Much less clearly do we realize that evolution is just as much a principle of *invariance*. Not only are those new anatomical conditions supposed to have originated according to the old laws of nature in general; it is also nothing but the primordial basic processes of the physical world which, according to the principle of evolution, are still operating under the new conditions. It follows that no elementary functional structure can appear in organisms that is not derivable from the fundamental concepts of science. And this applies also to the structure of value. Now, so long as science declares that nature before evolution was devoid of value in any sense, no amount of new organs can give us an understanding of value in terms of natural science. So much we have seen in a preceding discussion. Therefore, since the structural characteristics of value are said to have been absent from nature to begin with, and since, with this premise, evolution as a principle of change cannot be made responsible for those characteristics, we here find natural science contradicting itself. It adheres to evolution, i.e., to continuity and uniformity of all functions in the world. But it also denies any possibility of applying this principle in the case of value. We will next consider how science could extricate itself from this inconsistency.

It would, of course, be possible to sacrifice the principle of evolution. Then all difficulties would disappear excepting the one that the resulting dualism between neutral facts and value would destroy all hope that the unit of knowledge, of which modern philosophers of science are so rightly fond, will ever be achieved. I cannot believe that those philosophers or any scientists will be prepared to make such a sacrifice. But under these circumstances there is only one way out. Since, as a typical experience, value with its characteristic structure cannot be eliminated, and since science postulates the unity of all functional structures, science will have to deliberate whether its own concepts do not admit of certain modifications. For if science discarded the thesis that the functional characteristics of value are absent from nature, the inconsistency would disappear, the unity of knowledge would potentially be established, and evolution would become a principle which makes sense.

It is surprising to see how little such a step would affect science. There are two ways in which the necessary change can be discovered. We can either compare value situations with physical situations in general; or we can apply the same procedure to value situations and such particular situations in nature as are directly related to value experiences, namely, the somatic correlates of value situations in the brain of man. I prefer the second alternative. It will no longer be denied that any perceptual object is somatically represented by a cortical process, and that in each case this process is located in some particular part of the tissue. Similarly, the self in the sense in which we have used the term is represented by a complex of processes. We have every reason to assume that the location of this latter complex differs from that of a process which corresponds to an object. Now, in the case of a value situation, the percept in question has qualities which make it valuable. These qualities as such we have decided not to discuss. But we have found that no such quality alone constitutes actual value, because all objects which are valuable are sources of positive or negative demands. For the most part these are demands upon the self, which they either attract or repel. If we ask ourselves what somatic facts in the tissue could be the correlates of such demands, we are at once reminded of the physical concept "field." The field between two physical objects or processes extends from one to the other, and tends either to reduce or to widen their distance. The parallel is obvious. It is true, physicists vacillate between two interpretations of fields. Before Faraday fields were regarded as mere mathematical constructions; according to Faraday and Maxwell fields were the most important entities in nature; but at the present time there seems to be a certain reaction against this view, probably because the concept of field can hardly be divorced from that of causation, which is now unpopular. If we wish to give the demands of valuable percepts physical counterparts in the brain, we need do no more than adopt Faraday's view. Not only does a field attract an entity toward, or repel it from, the source of the field—just as the self is affected by the demand of a value. That attraction or repulsion in a field is also independent of further conditions which, apart from the field, may be operating in the given situation. It has sometimes been said that we find nowhere in nature an analogue of the difference between "happens" and "is," on the one hand, and "ought," on the other hand. This is incorrect since any object in nature that is subjected to a field remains thereby directed toward, or away from, the source of the field, even when, because of further conditions, it may actually move in another direction.

We have still to decide what the somatic counterpart of another characteristic of value is: the demand of a value object is felt to *follow from* the nature of the object. Now we have just seen that it depends merely upon our choice between several, otherwise equally acceptable, interpretations of the concept field whether or not a value demand can be referred to a somatic correlate. Our situation is precisely the same when we turn to the understandable relation between the nature of value objects and their demands. According to natural science the relation between objects and their fields is a relation of mere fact. But this proves only that the procedures of science have access to no more than such a factual connection. Whether the relation is actually of this nondescript type, or whether, if more were known, the nature of the field would in each case prove to follow from the nature of the objects involved—this is a question which natural science alone cannot answer. The issue simply lies beyond the possibilities of observation in physics. And, instead of excluding the latter alternative, science ought merely to say: I am unable to decide. We are therefore again at liberty to choose the interpretation which appears on the whole more satisfactory. In doing so we can never come into conflict with scientific data, because these data are not affected by our decision. Obviously, our choice will be such as to do justice to the nature of value, to establish the unity of knowledge, and to abolish the inconsistency in which science is now caught. This means that, on the evidence of phenomenology, we shall interpret the relation of objects and their fields as one of *following from*. With this choice, value situations as experienced and corresponding functional situations in the brain of man become structurally congruent.

It takes some courage to suggest that phenomenological observation be here accepted as evidence in matters of nature. But once the principle of evolution is taken for granted, such an attitude appears perfectly natural when we deal with problems of functional structure. We readily admit that, as to quantitative exactness and in the establishment of strict factual laws, the observational techniques of science achieve incomparably more than can be gained from phenomenology. But when we ask to precisely what relations these laws refer, then natural science is at the end of its resources, and restricts itself to noncommittal terms which just cover its own data. This is our starting point. We try to relate the phenomenology of value situations with corresponding somatic situations, and find that a demand extends from a value object to the self, just as a field in the brain would issue from an object process and affect the somatic correlate of the self. So far natural science can have no cause for objections. If, then, science assumes an agnostic attitude with regard to the particular relation between objects and their fields, any other evidence which fills the empty space must be welcomed. It happens that phenomenology sees more of that relation than is implied in the terms of least content which science likes to use. Under these circumstances, why should one further insist on an agnostic attitude?

Not even the parsimonious tendency of science offers an excuse. For it is not an important parsimony that tries to keep the number of data in the limited realm of natural science at a minimum. The clear, unitary system of knowledge as a whole is far more essential, and any schism in this respect must be regarded as a most undesirable expense. There is, moreover, one expense which this systematic parsimony of human thought simply cannot afford, and that is self-contradiction. Only the present suggestion seems to be capable of avoiding that expense where the relation of fact and value is concerned.

A SELECTED, ANNOTATED BIBLIOGRAPHY IN VALUE CLARIFICATION

Abbey, D. S. *Valuing*. Chicago: Instructional Dynamics, 1973.
This educational kit with cassette, booklets, and discussion guides for a small group emphasizes the value process as an approach to problem solving.

Assagioli, R. *Psychosynthesis*. New York: Viking Press, 1965.

Assagioli, R. *The act of will*. New York: Viking Press, 1973.
Assagioli, founder of psychosynthesis (a "height" psychology of the super conscious, as opposed to "depth" psychology) popularized the use of guided daydreams and fantasy in therapy and human growth. In *The Act of Will* he presents the will as the center of the person's being, and develops a psychology of the will that promotes human growth and self-actualization.

Banet, A. G. (Ed.). *Creative psychotherapy: A source book*. La Jolla, Calif.: University Associates, 1976.

Banet, A. G. Therapeutic intervention and the perception of process. In J. W. Pfeiffer & J. E. Jones (Eds.), *The 1974 annual handbook for group facilitators*. La Jolla, Calif.: University Associates, 1974.

Bourke, V. J. *Will in Western thought*. New York: Sheed & Ward, 1964.
This book provides a historical background to the understanding of the will. The author clarifies the many uses of the term *will*, which has been variously defined as intellectual preference, rational appetite, volition, and dynamic power. The last chapter is a history of the psychology of the will. This is an in-depth book for serious students.

Curran, C. A. *Counseling and psychotherapy: The pursuit of values*. New York: Sheed & Ward, 1968.

Curran, C. A. *Religious values in counseling and psychotherapy*. New York: Sheed & Ward, 1969.
Of all the theorists in counseling skills, Curran best integrates the process of valuing into counseling and psychotherapy. He discusses the tendency of therapists to avoid dealing with values because they are afraid of getting into moral and theological issues that they are not competent to deal with or that they do not believe in.

Executive Council of the Episcopal Church. *A workshop on value clarification*. New York: Seabury Press, 1970.
This forty-seven page booklet outlines an introductory course in value clarification for lay leaders in the church.

Hall, B. P. *Value clarification as learning process: A guidebook*. New York: Paulist Press, 1973.
Michael Kenny and Maury Smith are consultant authors of this volume which is primarily a handbook of strategies for students, teachers, and professional trainers. There are forty-six exercises described in the Guidebook. Part I gives a general introduction to value clarification; Part III gives designs for conferences; and Part IV discusses classroom applications.

Hall, B. P. *Value clarification as learning process: A sourcebook*. New York: Paulist Press, 1973.
Michael Kenny and Maury Smith are consultant authors of this volume that primarily deals with value-clarification theory from a humanistic, psychological, and educational perspective. Some of the outstanding contributions of the book are Chapter Five, which deals with guilt as a limitation and how apparent negative values may be converted to positive values that may be celebrated, and Chapter Seven, which deals with the values of work and leisure.

Hall, B. P. *The development of consciousness: A confluent theory of values*. New York: Paulist Press, 1976.
Brian Hall has expanded the vision of value-clarification theory into consciousness and human development. He sees value clarification as a methodology to be used with other human development models, and he outlines the eight stages of the development of consciousness from a primitive level of survival to the highest level of human potential and convivial technology. Hall points out particular applications for the future of organization development and leadership training, which he thinks will have to take into consideration the levels of consciousness and integrated sets of skills for each level.

Hall, B., Hendrix, J., & Smith, M. *Becomings: The education to wonder series*. New York: Paulist Press, 1974.
The authors apply value clarification to a religious education program that provides guidebooks for teachers, parents, and clergy. Parents' involvement in the religious education of their children is stressed. The guidebooks provide background and many activities.

Hall, B. P., & Smith, M. *Value clarification as learning process: A handbook*. New York: Paulist Press, 1973.
In this volume Hall and Smith apply value clarification to religious education. Chapter Three is a description of the inductive group process of experiential learning; Chapter Six develops constructs of work, maintenance, and leisure; and Chapter Seven discusses how value clarification may be used in a parish setting.

Hartwell, M., & Hawkins, L. (with Simon, S. B.). *Value clarification: Friends and other people*. Arlington Heights, Ill.: Paxcom, 1973.
The written materials are part of a complete audio-visual mini-course in value clarification dealing with friends and friendships as it affects young teen-

agers. There is a general introduction and eleven strategies that are fully described. Some of the strategies can be adapted to adult participants.

Hawley, R. *Value exploration through role playing*. New York: Hart Publishing, 1974.

The author presents many helpful guidelines, formats, and examples of ways to use role playing with junior and senior high school students. The book contains numerous ideas for using role plays to focus on values and on decision-making skills. This is an excellent resource for teachers who are looking for the support and the know-how to use role playing in the classroom.

Hawley, R., & Hawley, I. *Human values in the classroom: A handbook for teachers*. New York: Hart Publishing, 1973.

The Hawleys view human values as survival skills. In this book, they outline many specific teaching techniques and classroom procedures to enhance the development of valuing skills.

Hawley, R., Simon, S. B., & Britton, D. *Composition for personal growth: Values clarification through writing*. New York: Hart Publishing, 1973.

This teacher's handbook has a wealth of ideas on ways to develop writing skills and self-literacy simultaneously. The authors present many adaptable activities that aid students in "reading the book within themselves" and then in reflecting these values in composition and/or discussion.

Kirschenbaum, H. Clarifying values clarification: Some theoretical issues and a review of research. *Group & Organization Studies*, 1976, *(1)*, 99-116.

Kirschenbaum, H. *Advanced value clarification*. La Jolla, Calif.: University Associates, 1977.

A handbook for trainers, counselors, and teachers, this book explores theory and current research; develops designs for workshops and classes; tells how value clarification can be built into school curricula; and includes a comprehensive, annotated bibliography of value clarification.

Kirschenbaum, H., & Simon, S. B. (Eds.). *Readings in values clarification*. Minneapolis, Minn.: Winston Press, 1973.

This is a collection of thirty-eight articles on value clarification, varying in length and importance. Most of the articles deal with applying value clarification to specific school subjects, to religious education, to family life, and to education in general. One of the most outstanding articles is Howard Kirschenbaum's "Beyond Values Clarification," in which he shares his experiences in using value clarification and discusses some present trends. Another significant article is a reprint of Carl Rogers', "Toward a Modern Approach to Values: The Valuing Process in the Mature Person." (This article appears as Chapter 12 of *Freedom to Learn*.) There is an extensive annotated bibliography devoted exclusively to value clarification.

Maslow, A. H. *Religions, values, and peak experiences*. New York: The Viking Press, 1964.

Maslow, A. H. *Toward a psychology of being (2nd ed.)*. New York: Van Nostrand Reinhold, 1968.

Maslow, A. H. (Ed.). *New knowledge in human values*. Chicago: Henry Regnery Company, 1970.

Maslow, A. H. *The farther reaches of human nature*. New York: The Viking Press, 1971.

Maslow has done more than any other theorist to legitimize the scientific study of religion and values. Because Maslow seeks to establish objective values as opposed to subjective values, he provides a needed balance and integration for the value-clarification theorist. His approach is to research the values of self-actualized persons and then to encourage others to adopt similar values.

May, R. *Man's search for himself*. New York: W. W. Norton, 1953.

May believes that man's current extensive anxiety and emptiness results from a loss of values. He postulates that man's loss of a sense of worth and dignity as a human being is a major source of modern distress. He holds that the degree of an individual's inner strength and integrity will depend on how much that person believes in the values he lives by. May sees the valuing process as part of the person's growth in self-awareness, maturity, freedom, and responsibility.

May, R. *Love and will*. New York: W. W. Norton, 1969.

May considers the importance of the will in relation to the person and love. Part II deals with the will in terms of intentionality, and Part III explores the relationship between love and will.

Raths, L. E., Harmin, M., & Simon, S. *Values and teaching*. Columbus, Ohio: Charles E. Merrill, 1966.

This is the "grandfather" book that initiated the current interest in the value-clarification approach. Part II presents a theory of values; Part III deals with the value-clarifying method, especially the clarifying response and the value sheet; and Part IV describes how to use the value theory, how to get started, and discusses research completed and needed. A "classic," worth reading and studying.

Rogers, C. R. *Freedom to learn*. Columbus, Ohio: Charles E. Merrill, 1969.

Rogers is a masterful phenomonologist and in this book he presents the valuing process from his perspective. He deals well with values in general but does not specify any approach, process, structure, or skills for dealing with values in particular. Value clarification fits easily with Rogers' approach to experiential learning; they complement each other.

Rokeach, M. *The nature of human values*. New York: The Free Press, 1973.

Rokeach, an outstanding researcher in the field of values, tends toward the objective side of the value continuum. He is more interested in what values a person has than in value clarification's approach to helping people discover and form their values. Part I of his book on human values is well worth studying and comparing with value-clarification theory. Part IV on inducing value changes through cognitive and behavioral change summarizes worthwhile research. Rokeach has also developed a value survey instrument in which a participant ranks eighteen terminal values and eighteen instrumental values.

Sax, S., & Hollander, S. *Reality games: Games people "ought to" play*. New York: Popular Library, 1972.

This book is a collection of "growth games" and activities that cover the whole range of self, interpersonal relations, transpersonal dimensions, and social or organization growth systems. Chapter Six of Part I "Games for Integration" presents value-clarification theory briefly and then lists a number of strategies developed in consultation with Merrill Harmin.

Simon, S. *I am lovable and capable*. Niles, Ill.: Argus Communications, 1973.

This is a thirty-four page booklet that gives Simon's modern allegory on put-downs and self-appreciation. Suggestions are offered for how the allegory may be used to develop an appreciation of the worth of self and others.

Simon, S. *Meeting yourself halfway*. Niles, Ill.: Argus Communications, 1974.

This workbook gives thirty-one value-clarification strategies for daily living. Simon suggests that the exercises be utilized one each day over a one-month period.

Simon, S. B., Howe, L., & Kirschenbaum, H. *Values clarification: A handbook of practical strategies for teachers and students*. New York: Hart Publishing, 1972.

Containing seventy-nine strategies plus additional suggestions for the use of many of the strategies, this handbook focuses on primary and secondary education in the classroom. Many of the strategies are easily adapted to adult groups and may be used in workshops.

Smith, M. (Ed.). *Retreat resources: Designs and strategies for spiritual growth* (3 vols.). New York: Paulist Press, 1974; 1975.

These three volumes contain examples of how value clarification may be adapted to the needs of various groups. In Volume One Smith and Bloss use value clarification with religious sisters in a "Community Building Spiritual Renewal." In Volume Two the authors use value clarification with an adult group and with married couples. In Volume Three Bloss and Tobin adapt value clarification for use with young adults in a modern retreat setting.

Van Caster, M. *Values catechetics*. New York: Newman Press, 1970.

Van Caster has taken Hegelian philosophy and applied it to the religious education situation in several books that he has written. In this book on religious education he focuses on values, attempting to take objective values and subjective values as thesis and antithesis which develop into a new synthesis for the person involved in the growth process.